I0484393

EPA/600/R-02/092
November 2002

Envirogen Propane Biostimulation Technology for In-Situ Treatment of MTBE-Contaminated Ground Water

Innovative Technology Evaluation Report

Prepared by:

Technical Project Manager

Ann Azadpour-Keeley
Subsurface Protection and Remediation Division
National Risk Management Research Laboratory
Office of Research and Development
U.S. Environmental Protection Agency
Ada, OK 74820

National Risk Management Research Laboratory
Office of Research and Development
U.S. Environmental Protection Agency
Cincinnati, Ohio 45268

NOTICE

The U.S. Environmental Protection Agency (EPA) through its Office of Research and Development funded the information described here by Ann Keeley, the EPA TPM and WAM for this demonstration, under contract 68-C-98-138 to ManTech Environmental Research Services Corp. and 68-C-00-179 to SAIC. It has been subjected to the Agency's peer and administrative review and has been approved for publication as an EPA document. Mention of trade names or commercial products does not constitute an endorsement or recommendation for use.

All research projects making conclusions or recommendations based on environmental data funded by the U.S. Environmental Protection Agency are required to participate in the Agency Quality Assurance Program. This project was conducted under an approved Quality Assurance Project Plan. The procedures specified in this plan were used without exception. Information on the plan and documentation of the quality assurance activities and results are available from the principal Investigator.

FOREWORD

The U.S. Environmental Agency (EPA) is charged by Congress with protecting the nation's land, air, and water resources. Under a mandate of national environmental laws, the Agency strives to formulate and implement actions leading to a compatible balance between human activities and the ability of natural systems to nurture life. To meet this mandate, EPA's research program is providing data and technical support for solving environmental problems today and building a science knowledge base necessary to manage our ecological resources wisely, understand how pollutants affect our health, and prevent or reduce environmental risks in the future.

The National Risk Management Research Laboratory is the Agency's center for investigation of technological and management approaches for reducing risks from threats to human health and the environment. The focus of the Laboratory's research program is on methods for the prevention and control of pollution to air, land, water and subsurface resources; protection of water quality in public water systems; remediation of contaminated sites and ground water; and prevention and control of indoor air pollution. The goal of this research effort is to catalyze development and implementation of innovative, cost-effective environmental technologies; develop scientific and engineering information needed by EPA to support regulatory and policy decisions; and provide technical support and information transfer to ensure effective implementation of environmental regulations and strategies.

The purpose of this publication is to present information that will assist decision-makers in evaluating an innovative remedial technology for application to cleanup of sites with contaminated ground water. This ITER, which has been produced as part of the Laboratory's strategic long-term research plan, describes the effectiveness and applicability of the propane biostimulation technology developed by Envirogen as a potential in-situ remedial alternative for the mineralization of MTBE from contaminated ground water. This technology was demonstrated and evaluated at the Naval Base Ventura County at Port Hueneme, California. Spatial and temporal data to evaluate the technology were collected from a dense network of in-situ monitoring points over a period in excess of 300 days. This comprehensive evaluation of the Envirogen technology demonstrated that its application at this site did not meet the State of California's treatability criteria.

> Stephen G. Schmelling, Acting Director
> Subsurface Protection and Remediation Division
> National Risk Management Research Laboratory

ABSTRACT

The primary objective of the Biostimulation Technology Evaluation was to determine if enhanced biodegradation was occurring in a ground-water Test Plot to a sufficient degree to reduce intrinsic methyl tertiary butyl ether (MTBE) to the State of California's treatability criteria of 5 μg/L or below. The project was carried out at the National Environmental Technology Test Site (NETTS) at the (NBVC) Naval Base Ventura County, Port Hueneme, California where a hydrocarbon release into ground water occurred between September 1984 and March 1985 involving approximately 4,000 gallons of leaded and 6,800 gallons of unleaded premium gasoline.

The geology at the site consists of unconsolidated sediments composed of sands, silts, clays and minor amounts of gravel and fill material. A shallow, perched, unconfined aquifer is the uppermost water-bearing unit. The water table is generally encountered at depths between 6 to 8 feet below ground surface (BGS), and has a saturated aquifer thickness of 16 to 18 feet.

The evaluation was carried out between June 2001 and March 2002 using Control and Test Plots and a cadre of primary and secondary analytes through 15 sampling events. The goals of the project were approached with the use of deuterated MTBE (d-MTBE) and ground-water tracers including bromide and iodide.

An analysis of intrinsic MTBE, deuterated MTBE, daughter products, and geochemical parameters demonstrated that the technology did not meet the State of California's treatability criteria.

TABLE OF CONTENTS

TABLES OF CONTENTS (Continued)

TABLES OF CONTENTS (Continued)

List of Figures

List of Tables

ACRONYMS, ABBREVIATIONS, AND SYMBOLS

ALSI	Analytical Laboratory Services, Inc.
ARAR	Applicable or relevant and appropriate requirement
BIPT	Bacterial injection point in the Test Plot
Br^-	Bromide ion
BGS	Below ground surface
BTEX	Benzene, toluene, ethylbenzene, and xylenes
CAA	Clean Air Act
CERCLA	Comprehensive Emergency Response, Compensation, and Liability Act
CFR	Code of Federal Regulations
CFU	Colony Forming Units
Cl^-	Chloride ion
CO_2	Carbon dioxide
COC	Chain-of-Custody
CPT	Cone Penetrometer Technology
CWA	Clean Water Act
DBPR	Disinfection By-Product Rule
DO	Dissolved oxygen
DOC	Dissolved organic carbon
DOE	Department of Energy
d-MTBE	Deuterated methyl tert-butyl ether
EPA	U.S. Environmental Protection Agency
ITER	Innovative Technology Evaluation Report
LCS/LCSD	Laboratory control samples and laboratory control sample duplicates
MCL/MCLG	Maximum Contaminant Level and Maximum Contaminant Level Goal
MDL	Method detection limit
µg/L	Microgram per liter
mg/L	Milligram per liter
MS/MSD	Matrix spike/matrix spike duplicate
MTBE	Methyl tert-butyl ether
NAAQS	National Ambient Air Quality Standard
NETTS	Department of Defense National Environmental Technology Test Site

ACRONYMS, ABBREVIATIONS, AND SYMBOLS (Continued)

NBVC   Naval Base Ventura County

NEX   Naval Exchange

NFESC   Naval Facilities Engineering Service Center

NRMRL   National Risk Management Research Laboratory

OIPC   Oxygen injection point in the Control Plot

OIPT   Oxygen injection point in the Test Plot

OSWER   Office of Solid Waste and Emergency Response

PIPT   Propane injection point in the Test Plot

PMO   Propane monooxygenase

POB   Propane oxidizing bacteria

ppm   Part per million

PQA   Pre-Quality Assurance Project Plan Agreement

QA   Quality assurance

QAPP   Quality assurance project plan

QC   Quality control

RCRA   Resource Conservation and Recovery Act

RRF   Relative response factor

RPD   Relative percent difference

SAIC   Science Applications International Corporation

SDWA   Safe Drinking Water Act

SPRD   Subsurface Protection and Remediation Division

STDEV   Standard Deviation

SVE   Soil vacuum extraction

SVOC   Semi-volatile organic compound

TBA   *tert*-butyl alcohol

TCE   Trichloroethene

TPM   Technical Project Manager

TOC   Total organic carbon

TSCA   Toxic Substance Control Act

TSA   Technical system audit

ACRONYMS, ABBREVIATIONS, AND SYMBOLS (Continued)

UCL   Upper confidence limit

VMP   Vapor monitoring point

VOA   Volatile organic analysis

VOC   Volatile organic compound

WAM   Work Assignment Manager

WQCB   Water Quality Control Board

WQS   Water quality standard

ACKNOWLEDGMENTS

This report was prepared for the U.S. Environmental Protection Agency (EPA) by Ann Keeley, the EPA Technical Project Manager and Work Assignment Manager for this demonstration, at the National Risk Management Research Laboratory (NRMRL) in Ada, Oklahoma. The technology evaluation process was a cooperative effort that involved personnel from the EPA Office of Research and Development (ORD), EPA Region 9, U.S. Navy, California Water Quality Control Board (WQCB), and Envirogen.

The extensive effort of the following personnel during this project is gratefully acknowledged:

- Fran Kremer, Annette Gatchett, Bob Olexsey, and Steve Schmelling of NRMRL and Arlene Kabei of Region 9 for the composition of an outstanding management team for the overall MTBE demonstration evaluation program;

- The NRMRL QA Managers Ann Vega and Steve Vandegrift for their crucial roles in association with the various aspects of the quality assurance and quality control of this demonstration;

- Drs. Carl Enfield, John Wilson, and Randall Ross for their technical advice;

- Peter Raftery as well as the WQCB management for their technical and administrative efforts in granting the project permits;

- ManTech, a SPRD contractor for performing various tasks including system installation, sampling execution, and laboratory analytical services; and

- SAIC, a NRMRL contractor for the development of the project QAPP.

Special thanks are offered to the employees at the U.S. Navy, Naval Facilities Engineering Service Center (NFESC) host site for their hospitality and assistance throughout this demonstration, especially, Ernie Lory, Dorothy Cannon, and James Osgood.

EXECUTIVE SUMMARY

The primary objective of the Biostimulation Technology Evaluation was to determine if biodegradation was occurring in a ground-water Test Plot to a sufficient degree to reduce intrinsic MTBE to the State of California's treatability criteria of 5 µg/L or below. The evaluation was carried out using Control and Test Plots and a cadre of primary and secondary analytes through 15 sampling events over a 38-week test period. An analysis of intrinsic MTBE, deuterated MTBE, daughter products, and geochemical parameters demonstrated that the technology did not meet the State of California's treatability criteria.

The National Environmental Technology Test Site (NETTS) at the (NBVC) Naval Base Ventura County, Port Hueneme, California is the site of a hydrocarbon release into ground water (Everett et al., 1998) between September 1984 and March 1985 involving, according to inventory records, approximately 4,000 gallons of leaded and 6,800 gallons of unleaded premium gasoline. The resulting ground-water plume consists of approximately 9 acres of BTEX and approximately 36 additional acres of methyl tertiary butyl ether (MTBE) contamination, extending approximately 4,500 feet downgradient from the site of the release. The Port Hueneme NETTS facility is located approximately 40 miles northwest of Los Angeles.

The geology at the site consists of unconsolidated sediments composed of sands, silts, clays and minor amounts of gravel and fill material. A shallow, perched, unconfined aquifer is the uppermost water-bearing unit. The shallow aquifer is comprised of three depositional units: an upper silty-sand, an underlying fine- to coarse grained sand and a basal clay layer. Based on CPT pushes, the upper silty-sand unit ranges between 8 to 10 feet thick and the underlying sand is approximately 12 to 15 feet thick. The water table is generally encountered at depths between 6 to 8 feet below ground surface (BGS), with seasonal fluctuations ranging between 1 and 2 feet, yielding a saturated aquifer thickness of 16 to 18 feet near the test area.

Methyl *tert*-butyl ether (MTBE) has become the most widely used automobile fuel oxygenate (Gullick and leChevallier, 2002). As a consequence of fuel spills and leaking storage tanks, MTBE has become a ubiquitous and recalcitrant ground-water contaminant (Pankow et al., 1997; Rice et al., 1995; Reuter et al., 1998).

In an attempt to demonstrate ground-water remedial alternatives for MTBE, the U.S. Environmental Protection Agency (EPA) and the U.S. Navy entered into a memorandum of understanding (MOU) to conduct a demonstration of a treatment technology for MTBE in ground water. Technology vendors were chosen through an open solicitation requesting proposals for processes to treat MTBE. Proposals were then selected using external and internal peer review. Envirogen was selected to demonstrate their propane biostimulation barrier technology as a mechanism to inhibit the migration of MTBE through ground water. The potential remedial action proposes the stimulation of cometabolism by the injection of oxygen and propane into the aquifer along with MTBE degrading bacteria.

Project objectives were addressed through the establishment of treatment and control plots, a network of conventional upgradient and downgradient monitoring points in the aquifer and vadose zone, and a ground-water tracer mixing and injection system. The treatment plot received the vendor's biostimulation technology consisting of oxygen, propane, and bacterial amendments. The control plot received only oxygen.

The goals of the project were multifaceted with the end result being the determination of the efficacy of using propane and/or oxygen biostimulation and bioaugmentation as a potential remedial alternative for the removal of MTBE from ground water. Achieving these objectives was approached with the use of deuterated MTBE (d-MTBE) and ground-water tracers including bromide and iodide. The ratios of ground-water tracers between downgradient transects were designed to provide evidence concerning the relative losses in MTBE concentrations resulting from dilution and degradation. Likewise, the use of d-MTBE ratios in downgradient transects served as a tracer of anthropogenic MTBE. More importantly, the use of d-MTBE was selected to provide evidence of biodegradation by the realization of d-MTBE daughter products.

Bromide was used in a preliminary study to determine the velocity as well as the distribution of ground-water flow, and the degree of communication between the tracer injection system and each of the downgradient monitoring locations. Bromide injection was started on February 1, 2001, and was stopped on February 28, 2001. Monitoring continued in order to observe the return of bromide to background concentrations.

Based on the results of the pre-demonstration bromide tracer study, the final project plan was developed concerning the application rate of conservative and non-conservative tracers from the injection wells, and called for 15 sampling events rather than the original 7 because it was determined that little ground-water flow was taking place in other than the bottom portion of the aquifer. Periodic samples were taken from the middle and upper monitoring screens, however, to assure that flow remained predominantly at the bottom of the aquifer through the evaluation period.

During the latter part of May 2001, the performance evaluation phase of the project was begun with the addition of amendments of oxygen, propane, and bacteria. The injection of iodide started on June 8, 2001. Iodide was selected for use in this phase of the project because of its low level of detection and to avoid possible problems associated with residual bromide concentrations. The first sampling event took place on June 14, 2001.

Some significant observations were made concerning the period during the pre-characterization investigation, beginning in late 2000, up to the beginning of the evaluation period in June 2001. For example, the overall intrinsic MTBE concentration in the vicinity of the plots dropped about 500 μg/L between October 4 and November 11, 2000, and MTBE concentrations in the Control Plot were significantly higher than those in the Test Plot. Most significantly, MTBE concentrations in the downgradient Test Plot dropped from over 5,000 μg/L in January 2001 to less than 1,000 μg/L by the first sampling event in June. This meant the remediation technology had to be effective in reducing the MTBE concentration from less than 1,000 μg/L to 5 μg/L or below rather than starting with a MTBE concentration of over 5,000 μg/L.

During the 38-week period between June 14, 2001, and March 8, 2002, 15 sampling events took place, occurring biweekly for the first ten events and monthly thereafter. Although sampling was concentrated at the bottom well screens, the middle and upper screens were sampled periodically at each well location. In the Test Plot the sampling locations included 6 upgradient wells, 14 downgradient wells, and 19 injection wells. The Control Plot consisted of 4 upgradient wells, 10 downgradient wells, and 19 injection wells.

In addition to the primary parameters of MTBE, d-MTBE, and iodide, samples were also analyzed for appropriate secondary parameters in order to test for both MTBE and d-MTBE daughter products as well as changes in geochemistry. Following the evaluation period it was determined that geochemical parameters in the upgradient and downgradient Test and Control Plots were unchanged. There was no evidence of increases in alkalinity in the downgradient Test Plot as would be expected, nutrients were not reduced, and most importantly, the total and dissolved organic carbon (electron donors) were not reduced.

The daughter products which were analyzed included: acetone; acetone-d6; 2-propanol; 2-propanol-d6,d8; formaldehyde; *tert*-butyl alcohol; and *tert*-butyl alcohol-d9,d10. Very low levels of daughter products were detected in both the Test and Control Plots. While only TBA was detected at the upgradient wells, both d-TBA and TBA were detected in the downgradient wells. It was not determined whether biotic or abiotic processes produced these products.

The d-MTBE in both the Test and Control Plots increased throughout the evaluation period. Although the concentrations were slightly higher in the Control Plot because of its higher hydraulic conductivity, the increase in both Plots was the same as determined by a least squares fit of the data.

The intrinsic MTBE concentrations in the upgradient Test Plot and both upgradient and downgradient portions of the control Plot decreased gradually through the evaluation period. In the downgradient Test Plot, the most significant site of the evaluation, the data remained between 300 – 600 µg/L with a small positive slope as determined by a least squares calculation.

SECTION 1

INTRODUCTION

The Envirogen propane biostimulation technology (the technology) was demonstrated for the treatment of ground water contaminated with methyl *tert*-butyl ether (MTBE) over a 300-day period at the Department of Defense National Environmental Technology Test Site (NETTS) at the Naval Base Ventura County (NBVC) at Port Hueneme, California. This Innovative Technology Evaluation Report (ITER) describes the results of that demonstration and provides other pertinent technical and cost information for potential users of this technology. For additional information about this technology, and the evaluation, refer to key contacts listed at the end of this section.

1.1   PURPOSES AND ORGANIZATION OF THE ITER

Information presented in the ITER is intended to assist decision makers in evaluating specific technologies for a particular cleanup situation. The ITER represents a critical step in the development and commercialization of a treatment technology. The report discusses the effectiveness and applicability of the technology and analyzes costs associated with its application. The technology's effectiveness is evaluated based on data collected during the demonstration. The applicability of the technology is discussed in terms of waste and site characteristics that could affect technology performance, material handling requirements, technology limitations, and other factors.

The purpose of this ITER is to present information that will assist decision makers in evaluating the Envirogen propane biostimulation technology for application to a particular site cleanup. This report provides background information and introduces the propane biostimulation technology (Section 1.0), provides an overview of demonstration objective and evaluation justification of the technology demonstration at the NBVC (Section 2.0), describes performance monitoring approach and sampling and analysis protocol (Sections 3.0 and 4.0), provides an overview of pre-demonstration tracer test (Section 5.0), analyzes the technology's applications (Sections 6.0 and 7.0), analyzes the economics of using the propane biostimulation technology to treat contaminated ground water (Section 8.0), summarizes the technology's applications analysis (Section 9.0), describes the technology's status (Section 10), and presents a list of references used to prepare the ITER. Vendor's claims for the propane biostimulation technology are presented in Appendix A.

1.2 DESCRIPTION OF THE MTBE DEMONSTRATION PROGRAM

In 1999, the U.S. Environmental Protection Agency (EPA) and the U.S. Navy entered into a memorandum of understanding to conduct a multi-year program involving demonstration and evaluation of several innovative technologies for the treatment of MTBE in ground water. Technology vendors were identified through an open solicitation requesting proposals for processes to treat MTBE. Vendors participating in the program were selected based on the results of external and internal EPA/Navy peer review processes.

The site that was selected through an open solicitation to host the multiple-vendor MTBE demonstration program was the Department of Defense National Environmental Technology Test Site (NETTS) at the Naval Base Ventura County (NBVC) at Port Hueneme, California. The Port Hueneme NETTS facility is located approximately 40 miles northwest of Los Angeles. The Naval Exchange (NEX) service station is the source of the petroleum plume that occurs at the Port Hueneme NBVC facility. The NEX service station site is typical of similar gasoline service station sites throughout the country, where leaking gasoline storage tanks and product delivery lines have contaminated surrounding ground water with gasoline compounds and additives, including MTBE (Kostecki et al., 1997). According to NEX inventory records, approximately 4,000 gallons of leaded and 6,800 gallons of unleaded premium gasoline were released from the distribution lines between September 1984 and March 1985. The MTBE plume that emanates from the NEX service station at the NBVC site extends approximately 4,500 feet from the contamination source in a shallow perched aquifer.

Three locations within the MTBE plume at the NEX service station site were identified as potential locations for technology demonstrations. These three locations are differentiated by their distance from the source and are identified as follows:

1. **The Source Zone**: This zone is located within the immediate vicinity of the source and is characterized by having a high concentration of MTBE as well as benzene, toluene, ethylbenzene, and xylenes (BTEX), and potentially contains free-phase gasoline.

2. **The Middle Zone**: This zone is the area mid-way downgradient along the MTBE plume and contains moderate concentrations of MTBE.

3. **The Wellhead Protection Zone**: This zone is farthest downgradient along the plume, and contains MTBE at lower concentrations than the first two zones.

Figure 1-1 indicates the extent of the MTBE plume at Port Hueneme as of August 1999, and identifies the three zones within the plume; Figure 1-2 provides an expanded view of the Middle Zone, the location of the Envirogen technology demonstration.

1.3   TECHNOLOGY DESCRIPTION

This section describes the Envirogen propane biostimulation technology that was demonstrated at the NBVC, Port Hueneme, California.

1.3.1   Principles of the Propane Biostimulation Technology

The Envirogen technology that was applied in this demonstration was an extension of conventional biosparging methodologies in that pure oxygen and propane sparging were applied in a biostimulation mode. The conceptual approach involved the addition of oxygen (for aerobic respiration) and propane (as a cosubstrate) to stimulate the propane oxidizing bacteria (POB) in the production of the enzyme propane monooxygenase (PMO) that catalyzes the degradation of MTBE and its primary degradation product, TBA, to carbon dioxide and water (Figure 1-3). Exogenous propane oxidizing bacteria (POB) *Rhodococcus ruber* strain ENV425 was used to seed the aquifer at the onset of the demonstration to insure activity and speed initiation of the treatment process.

Envirogen claims that oxygen and propane flow rates were designed to provide an adequate substrate to create an aerobic treatment zone and stimulate enzyme production, while minimizing the stripping of VOCs and off-gassing propane and oxygen. Therefore, much lower oxygen injection flow rates were required for their process compared to conventional biosparging. Gases can be injected into conventional sparging wells, using permeable membranes or tubing, or using in-well sparging or mixing techniques. Because substrate mixing occurs within the saturated aquifer, soil vacuum extraction (SVE) operation is typically not required.

A review of the technical literature suggests that the biostimulation technologies can be applied in a variety of configurations to provide source area treatment or downgradient plume containment, depending on site characteristics and remediation needs including:

1.   A modified multi-point air sparging system (Salanitro et al., 1999 and 2000; Benner et al., 2000; Clayton et al., 1995; Ji et al., 1993; Johnson, 1994; Johnson et al., 1993 and 1996; Pankow et al., 1993) that delivers propane air or oxygen throughout a contaminated site (suitable for use with existing systems or specially designed systems),

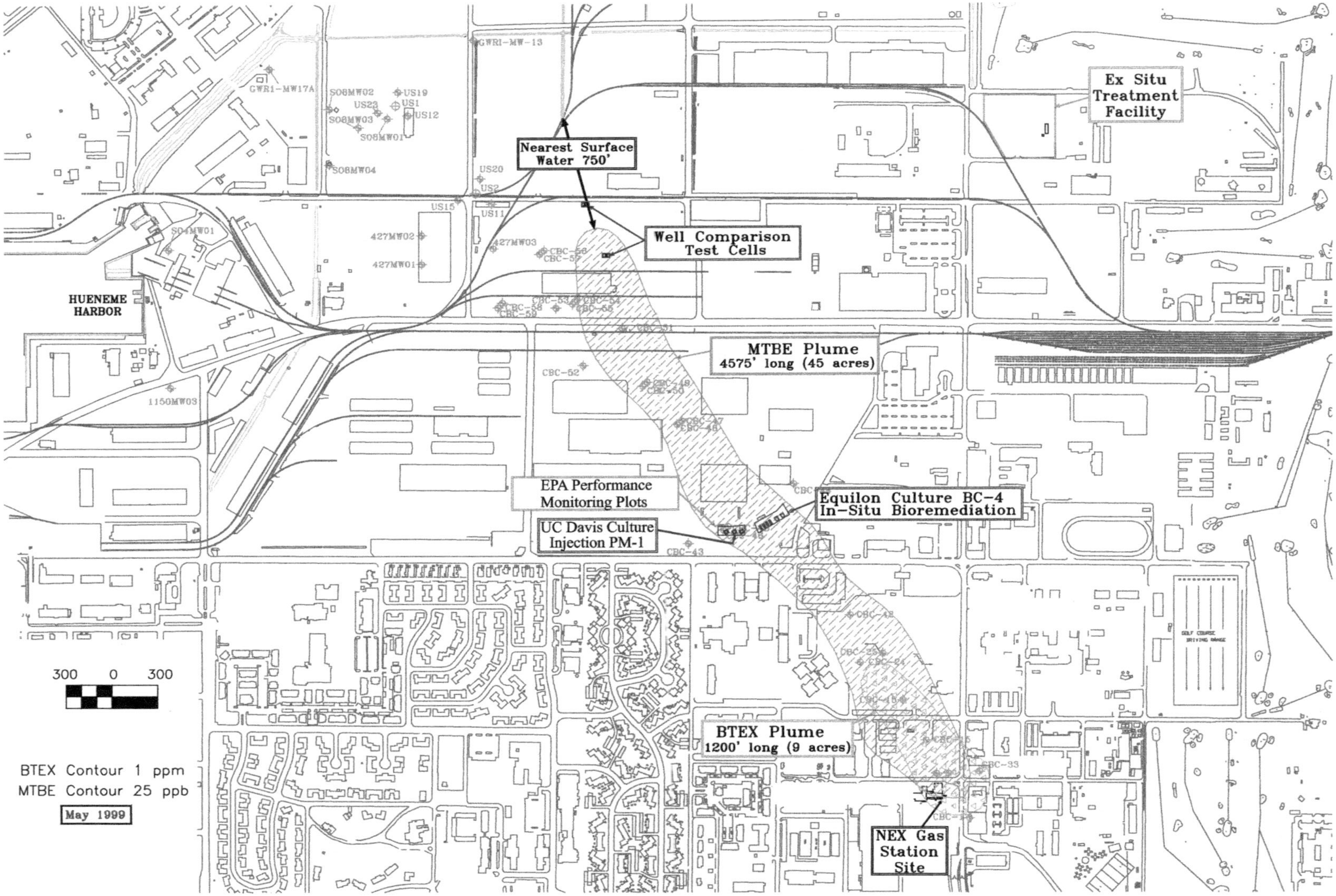

Figure 1-1. Port Hueneme Plume Map

BTEX Contour 1 ppm
MTBE Contour 25 ppb

May 1999

300   0   300

Ex Situ Treatment Facility

Well Comparison Test Cells

MTBE Plume 4575' long (45 acres)

Equilon Culture BC-4 In-Situ Bioremediation

Nearest Surface Water 750'

EPA Performance Monitoring Plots

UC Davis Culture Injection PM-1

BTEX Plume 1200' long (9 acres)

NEX Gas Station Site

HUENEME HARBOR

8

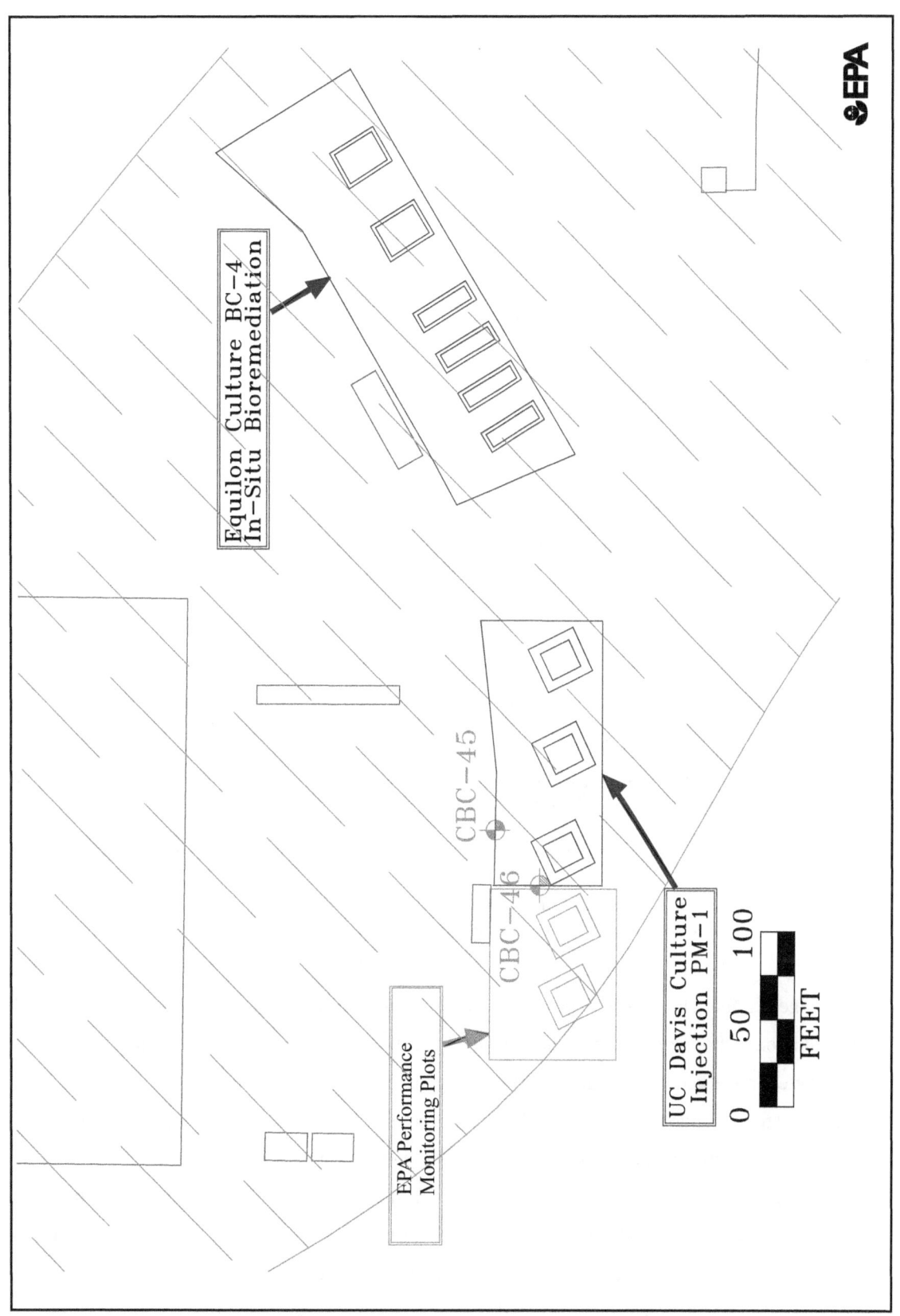

Figure 1-2. Site Location

9

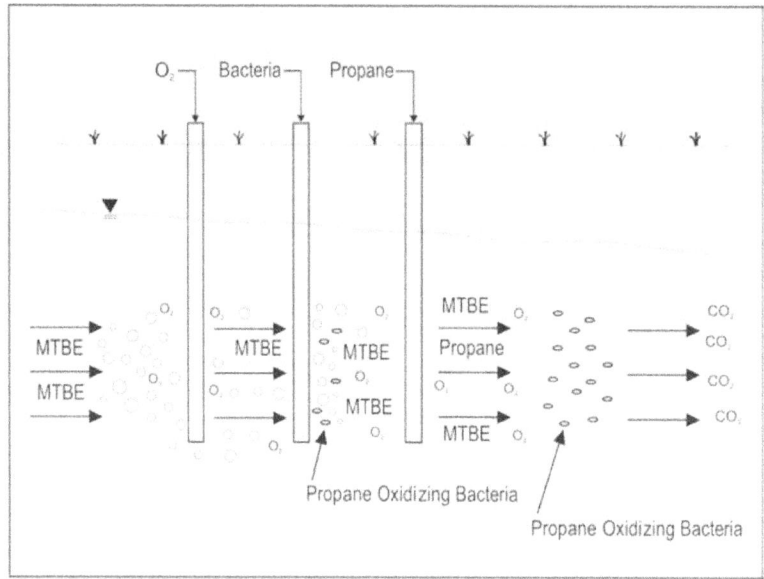

Figure 1-3. In-Situ Application of Propane Biostimulation.

(Modified from *Draft In-Situ Remediation of MTBE Contaminated Aquifers Using Propane Biostimulation, Technology Demonstration* Plan, 2000, by Envirogen.)

(A variety of configurations to apply propane biostimulation technology continued from Page 7):

2.   A series of oxygen/propane delivery points arranged to form a permeable treatment wall to prevent off site migration of MTBE,

3.   A permeable treatment trench fitted with oxygen and propane injection systems, and

4.   An in-situ recirculating treatment cell that relies on pumping and reinjection to capture and treat a migrating contaminant plume.

Envirogen further claims that propane biostimulation has several advantages over existing MTBE remediation technologies. The primary advantage is that the technology can be applied in-situ to completely remediate MTBE and TBA without generation of waste products. Because propane biostimulation technology is an extension of conventional air sparging and biosparging techniques, the existing knowledge base regarding their design and implementation allows simplified application of the technology. Moreover, the addition of propane injection to existing or new systems may be able to be accomplished with minimal added equipment and costs. Because the technology is complementary to air sparging, biostimulation treatment zones can be developed in conjunction with source treatment measures to address BTEX and other fuel hydrocarbons. If inhibition arises due to the presence of these compounds, the propane biostimulation treatment zone can be established downgradient by adding additional oxygen and propane treatment, and applied sequentially after BTEX compound concentrations are reduced.

For this demonstration, the propane and oxygen were injected into the saturated aquifer using sparging wells and pressurized gas systems designed to provide flexible performance characteristics and safe operation. Oxygen and propane were intermittently sparged into the aquifer using separate sparge points at a total rate of approximately 1- to 10-pounds/day and 0.1- to 0.5- pounds/ day, respectively. The vendor (Envirogen) claims that the frequency and duration of sparging were optimized based on the results from start-up monitoring to minimize off gassing of oxygen and propane to the vadose zone and stripping of MTBE.

1.3.2   Demonstration System Design

The Envirogen's demonstration system consisted of a network of oxygen, bacteria, and propane injection points, pressurized oxygen and propane gas delivery and control systems, and ground-water and soil-gas monitoring network. Figure 3-1 illustrates the layout of the demonstration system.

The Test Plot consisted of a network of injection wells designed to deliver oxygen and propane into the ground water to stimulate the appropriate microbial metabolic processes. Eight-oxygen injection points (OIPTs) and 7-propane injection points (PIPTs) were installed as shown on Figure 3-1. The OIPTs were spaced 1-meter (3.28 feet) apart on a line perpendicular to ground-water flow. The PIPTs were placed approximately 1.5 meters (4.9 feet) downgradient of the OIPTs, and offset from the OIPTs. Eight-bacterial injection points (BIPT) were installed between OIPTs and PIPTs. The BIPT were used for a one time release of bacterial suspensions that occurred 16 days after the start of oxygen injection. On May 23, 2001, the equivalent of 5 liters of bacterial culture at a density of 10^{11} cfu/mL were injected into the aquifer at the Port Hueneme site, resulting in a final bacterial density in the aquifer of approximately 10^8 cfu/mL.

The Test Plot ground-water monitoring network consisted of 16 dual-level, nested wells. A background well nest was placed along the centerline of the OIPTs, approximately 2.5 meters (8.2 feet) upgradient of the OIPTs. Envirogen wells were placed in 5-rows of two nested wells each, at downgradient distances from the PIPTs. The center well in each row was aligned with the centerline of the OIPTs. The soil-gas monitoring network consisted of 6 vapor monitoring points (VMPs) distributed around the OIPTs and PIPTs.

The Control Plot was similar in configuration to the Test Plot, except that no bacteria or propane injection points and fewer monitoring points were installed. Figure 3-1 illustrates the Control Plot configuration. Eight-OIPCs were installed at 1-meter (3.28 feet) spacing along a line-oriented perpendicular to ground-water flow. The ground-water monitoring network consisted of 10 dual-level, nested wells: 1-upgradient well nest and 3-rows of performance monitoring wells, placed at 2.25, 4 and 6.5 meters (7.4, 13.1, and 21.3 feet) downgradient from the OIPCs. The soil-gas monitoring network consisted of 4-VMPs placed around the OIPCs.

Oxygen and propane injection points were installed using Geoprobe™ methods. The OIPTs, OIPCs , and PIPTs were installed through the direct push rods using an expendable tip to anchor the assembly in the formation at the design depth. Oxygen and propane injection points were constructed using 1-inch ID, Schedule 40 PVC casings from 2-feet above the ground surface to approximately 10-feet below the water table. The well screens were constructed using 1-foot length Schumaprobe ™ screens composed of sintered polyethylene. The design for the propane and oxygen biostimulation system was based on the anticipated requirements associated with a relatively small area. As such, the equipment required to provide and control oxygen and propane supply were portable. The system consisted of pressurized oxygen and propane tanks, individual oxygen and propane control manifold assemblies and a control panel equipped with timers to allow pulsed operation of the injection systems. Figure 1-4 illustrates the piping and instrumentation diagram for the system.

Separate oxygen distribution systems were set-up for the Test and Control Plots. Each plot utilized two 75-pound oxygen cylinders piped in series with appropriate pressure regulators to allow oxygen delivery at approximately 60 pounds per square inch gage (PSIG). Oxygen flow to the manifold was controlled using a timer actuated solenoid valve. Flow and operating pressure at each oxygen injection point well-head was controlled using individual needle valves, sized to allow oxygen flow rates of 1 to 60 standard cubic feet per hour (SCFH) at operating pressures of up to 12 PSIG. Each wellhead was equipped with a flow meter and pressure valve port to allow flow balancing and system performance monitoring. The primary distribution line from the oxygen tanks, manifold assembly and individual wellhead distribution laterals were constructed of materials designed for oxygen. The oxygen tanks for the Control and Test Plots were housed in separate cages located near each plot as shown on Figure 1-4.

The Test Plot propane distribution system consisted of one 35-pound propane cylinder with appropriate pressure regulator to allow propane delivery at 30 PSIG. Propane flow to the manifold assembly was controlled using a timer actuated solenoid valve. Flow and operating pressure at each propane injection point wellhead was controlled using individual needle valves, sized to allow propane flow rates of 0.5 to 5 SCFH at 12 PSIG. Each wellhead was equipped with a dedicated flow meter and pressure valve port to allow flow balancing and system performance monitoring. The primary distribution line from the propane tank, manifold assembly and individual wellhead distribution laterals were constructed of materials specifically designed for propane. The propane tank was housed in a separate cage near the Test Plot.

The control panel was mounted on the exterior wall of the EPA shed in proximity to the utility pole and properly anchored and grounded. The demonstration system utilized 3-phase, 208V power supplied by NETTS. The system controls operated using conditioned power reduced to 24V AC power to the individual timers and solenoid valves. The system was fabricated and shipped to the demonstration site to meet the demonstration startup. The individual system components were pre-assembled in a modular fashion for ease of shipping and field assembly.

1.4   KEY CONTACTS

Additional information about the propane biostimulation technology and the NBVC demonstration can be obtained from the following sources:

Ann Keeley, Ph.D.
U.S. Environmental Protection Agency
Office of Research and Development
National Risk Management Research Laboratory
Subsurface Protection and Remediation Division
919 Kerr Research Drive
Ada, Oklahoma 74820
Telephone: (580) 436-8890
FAX: (580) 436- 8614
Email: keeley.ann@epa.gov

Rob Steffan, Ph.D.
Envirogen
4100 Quakerbridge Road
Lawrenceville, NJ 08648
Telephone: (609) 936-6300
FAX: (609) 936-9221
Email: steffan@envirogen.com

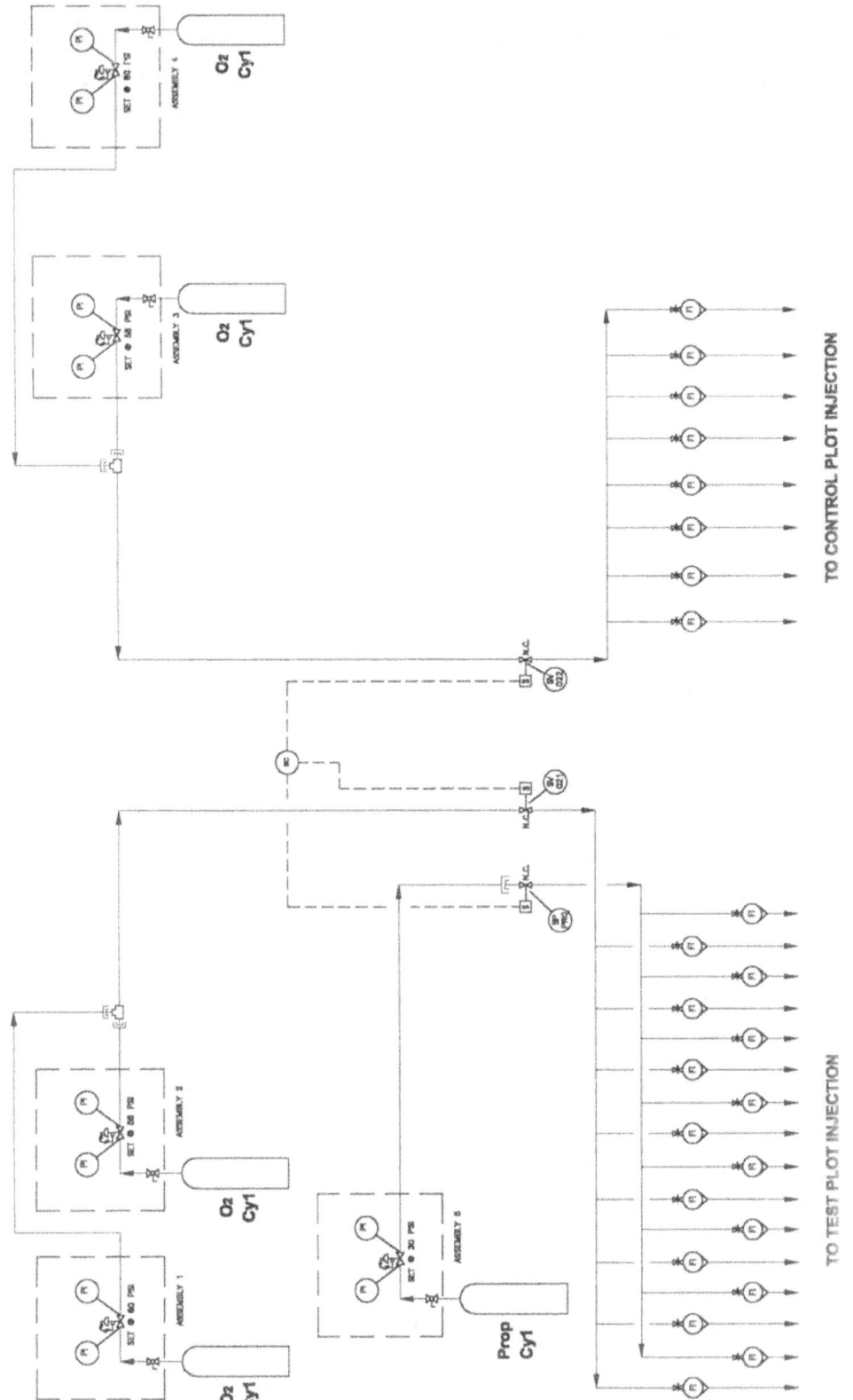

Figure 1-4. Piping Instrumentation Diagram

(Adopted from Draft *In-Situ Remediation of MTBE Contaminated Aquifers Using Propane Biostimulation, Technology Demonstration Plan*, 2000, by Envirogen.)

14

SECTION 2

DEMONSTRATION OBJECTIVE AND EVALUATION JUSTIFICATION

The overall objective of the MTBE demonstration program was to exhibit and evaluate the cost and performance of innovative MTBE treatment technologies at MTBE source and wellhead areas. Unequivocally, regardless of the technology type or its application within the source, middle, or wellhead zone, the final selection for the treatability criteria was to achieve the California target cleanup goal of 5 μg/L. Accordingly, each technology was evaluated on its own merit and was not compared to any other technology.

The final selection of various technologies upon their acceptance of meeting the cleanup goal of 5 μg/L at the NBVC was guided by the technical representatives of a group of stakeholders that included the following organizations:

> U.S. EPA, National Risk Management Research Laboratory (NRMRL)
>
> U.S. Navy, Naval Facilities Engineering Service Center (NFESC)
>
> U.S. EPA, Region 9
>
> California Department of Health Services (DHS)
>
> California Water Quality Control Board (WQCB)

Each of the stakeholders participated in conference calls and a meeting at the site in September 2000, to discuss the technical details of the demonstration and to assure that the technical approach to the demonstration adequately addressed elements of interest to potential users of each technology.

2.1   BACKGROUND

The ground water at the demonstration location within the Middle Zone was known to be contaminated with gasoline constituents. The primary components of environmental concern included MTBE, and products of MTBE degradation, including primarily *tert*-butyl alcohol (TBA). To confirm the presence of these components and their approximate concentrations in the ground water in the vicinity of the proposed test plots, an overview of the host site characterization is described.

2.1.1   NBVC Site Characterization

This section provides information gathered at the Port Hueneme site prior to the detailed design of the in-situ MTBE bioremediation performance monitoring work plan and QAPP. Essentially the pre-demonstration characterization consisted of a determination of the background concentrations of chemicals of interest to the demonstration.

2.1.2   Site Description

The selected location (site) for the propane biostimulation technology demonstration is situated approximately 2,400 feet southwest of the NEX station, adjacent to the U.C. Davis and Equilon demonstration plots (Figure 1-2). The area surrounding Envirogen's biostimulation technology demonstration site has been characterized during prior site investigation activities and includes 4 monitoring wells (CBC-43, CBC-44, CBC-45 and CBC-46) and 9 cone penetrometer (CPT) pushes to determine stratigraphy of the soil.

2.1.3   Hydrogeology

The geology at the site consists of unconsolidated sediments composed of sands, silts, clays and minor amounts of gravel and fill material. A shallow, perched, unconfined aquifer is the uppermost water-bearing unit. The shallow aquifer is comprised of three depositional units: an upper silty-sand, an underlying fine- to course grained sand and a basal clay layer. Based on CPT pushes, the upper silty-sand unit ranges between 8 to 10 feet thick and the underlying sand is approximately 12 to 15 feet thick. The water table is generally encountered at depths between 6 to 8 feet below ground surface (BGS), with seasonal fluctuations ranging between 1 and 2 feet, yielding a saturated aquifer thickness of 16 to 18 feet near the test area.

Ground-water flow is generally to the southwest under hydraulic gradients between 0.001 and 0.003 ft/ft. Transmissivity estimates for the shallow aquifer were derived based on pumping tests and slug tests, with results ranging between 2,500 and 6,500 ft^2/day. Based on an average saturated thickness of 17 feet, hydraulic conductivity estimates range between 170 and 440 ft/day (6×10^{-2} to 1.6×10^{-1} cm/s). Estimated ground-water flow velocity ranges between 230 and 330 feet/year, assuming an aquifer porosity of 0.35.

2.1.4 Contaminant Distribution

Ground-water contamination is limited to the perched aquifer across the NBVC facility. Monitoring wells CBC-45 and CBC-46 represent the ground-water quality conditions within the dissolved MTBE plume near the demonstration site (Figure 1-2). Historical ground-water sampling from these wells between September 1998 and September 1999 indicate MTBE concentrations ranging between 6,300 to 3,500 µg/L at CBC-45 and 4,000 to 1,100 µg/L at CBC-46. (CBC-45 is screened at 17' - 18' BGS and CBC-46 is screened at 12' - 13' BGS.) Apart from a *tert*-butyl alcohol (TBA), an intermediate metabolite, detection of 470 µg/L at CBC-45 in June 1999, none of the other samples exhibited TBA or gasoline (BTEX compounds) concentrations above their respective practical analytical laboratory quantification limits.

Based on the initial hydrogeologic and contaminant characteristics, it was believed that the selected site would provide an ideal setting for evaluating the performance of the propane biostimulation approach. In order to confirm the presence of the primary components of environmental concern at the proposed Envirogen demonstration site, EPA's Subsurface Protection and Remediation Division (SPRD), Ada, Oklahoma, performed pre-characterization activities as described in the following section.

2.2   TECHNOLOGY DEMONSTRATION SITE PRE-CHARACTERIZATION

To characterize the contaminated ground water at the demonstration location within the Middle Zone, on September 26, 2000, during EPA's pre-characterization activities at Port Hueneme, ten monitoring wells were installed inside the in-situ bioremediation (Envirogen) fenced area. The ten 2-inch wells were constructed and developed using a direct push rig, 5 wells upgradient, in a straight line immediately inside the east fence of the site (wells ID 6 - 10) and 5 wells downgradient, in a straight line immediately inside the west fence (wells ID 1 - 5). The two lines are approximately 50 feet apart.

Wells ID 2 - 5 were sampled twice, once with filtration and once without filtration. First, water samples ID 1 - 5 were collected after passing through the water filter (water flow-through cell); and, second, ID samples 2 - 5 were re-sampled directly from the wells. Samples from wells ID 6 - 10 were collected directly from the wells without filtration. The sampling direction was south to north (assuming an increasing concentration gradient). The samples were analyzed by CapCo Analytical Services Inc., Ventura, California, for MTBE and d-MTBE (Practical Quantitation Limit of 5.0 µg/L), TBA, and BTEX

17

by EPA Method 8260 (EPA 1995; 1996). The results indicated there was no d-MTBE or BTEX, while MTBE ranged from 296 - 5,040 µg/L. The wells were sampled again on October 4, 2000, and were analyzed by ManTech Environmental Services (ManTech) in Ada, Oklahoma, according to RSKSOP-217 "Determination of Methyl Tertiary Butyl Ether, Tertiary Butyl Alcohol, and Volatile Aromatic Compounds in Water by Automated Headspace Gas Chromatography/Mass Spectrometry." The results of the laboratory analysis of these ground-water samples are shown in Table 2-1 and confirmed the presence of the expected gasoline components. Based on the results of these analyses, the study participants concluded that the proposed test area contained adequate levels of MTBE in the ground water to challenge Envirogen's biological barrier technology. Non-detectable levels of BTEX compounds were also confirmed for the process.

During the month of November 2000, additional pre-characterization samples were obtained from the site for purposes of method proficiency testing. These samples were analyzed by 4 different laboratories in order to confirm that MTBE analyses could be performed by conventional purge-and-trap procedures. Summarized in Table 2-1 are data from this study to further support the supposition that ground water at the site contain adequate concentrations of MTBE and also show that no BTEX is present.

In addition to the gasoline components identified above, the stakeholders identified a number of potential by-products of biochemical oxidation that may well be formed during treatment of ground water using the Envirogen technology. Specifically, by-products from the microbial oxidation of MTBE were expected to include TBA, acetone, 2-propanol, and formaldehyde as shown in Table 2-2.

The contaminants of interest identified above were, therefore, included on the list of parameters to be determined in both upgradient and downgradient samples during the demonstration in order to assess the effectiveness of the Envirogen treatment. Based on the review of the regulatory criteria for these contaminants of interest and discussions among the stakeholders, treatment goals were established for selected contaminants of interest as listed in Table 2-3. The treatment goals for MTBE and TBA were identified as the lowest maximum contaminant level (MCL) or action level (AL) promulgated by the State of California. No treatment goal was set for the Disinfection By-Product Rule (DBPR) since ground water was not to be used as a drinking water supply. The other regulatory criteria presented in Table 2-3 for critical and non-critical parameters were used as advisory information and not as basis for setting the treatment goals for the Envirogen demonstration.

The demonstration of the Envirogen technology was implemented in one phase, the technology was evaluated over a ten-month period during June 2001 to March 2002 at presumably steady-state operating conditions. For this demonstration, a specific set of objectives was formulated and a quality assurance project plan (QAPP) was written to guide the EPA field sampling, laboratory analysis, and data evaluation efforts.

2.3   DEMONSTRATION OBJECTIVES

The objectives of the demonstration were to determine the effectiveness of using propane and oxygen biostimulation and bioaugmentation [exogenous propane oxidizing bacteria (POB) *Rhodococcus ruber* strain ENV425] as a potential remedial alternative for the removal of MTBE from ground water. Although MTBE concentrations upgradient and downgradient of the treatment system were evaluated, the technology critique was centered on the fate of d-MTBE added to the system. The d-MTBE and iodide (a conservative ground-water tracer) were used to evaluate biotic and abiotic (dispersion) attenuation as the contaminant passed through the biological barrier. The ratio of the ground-water tracers between downgradient transects provided evidence concerning the relative losses of MTBE resulting from dispersion and degradation. The use of d-MTBE provided evidence of biodegradation by tracking the generation of d-MTBE daughter products. For this demonstration, the deuterated daughter products that were tracked include d-TBA, d-2-propanol, and d-acetone. To meet the specific project requirements, the determination of MTBE, d-MTBE, and their respective metabolites were accomplished by the analysis of collected samples using GC/MS methodology with reporting limits (minimum quantitation limits) that were, at the minimum, 100 times lower than the applied concentration of the tracers of interest.

Project objectives were met through the establishment of Test and Control Plots, a network of conventional upgradient and downgradient monitoring points in the aquifer and vadose zone, and a ground-water tracer mixing and injection system. The treatment plot received Envirogen's biostimulation technology consisting of oxygen, propane, and POB amendments. The Control Plot received only oxygen amendments. A ground-water tracer injection system was used to determine spatial ground-water flow patterns (vertical and horizontal) in the Test and Control Plots before and during treatment.

The evaluation of the Envirogen demonstration consisted of two phases including the pre-demonstration tracer study and the long-term demonstration evaluation. During the first phase, bromide was used to

Characterize ground-water flow and time of travel. The second phase incorporated the use of d-MTBE and iodide to address expected decreases in d-MTBE concentrations and determine the extent of the reductions with respect to biodegradation or dilution.

Also, as noted above, a critical phase of this technology performance evaluation was a ground-water tracer study utilizing a mixing and injection system. This system, designed and installed by SPRD delivered bromide into the ground water at both the Test and Control Plots during the pre-demonstration period to achieve two objectives. First, to validate the hydraulic properties at the site, and second, to determine the efficacy of the tracer system for its use during the long-term performance demonstration period.

During the pre-demonstration activities consisting of approximately 4 weeks of bromide injection (January 31 – February 28) and subsequent on-going monitoring, the system operated adequately to allow an evaluation of the effectiveness of the demonstration. This was supported by significant bromide "hits"(concentration of approximately 2 ppm and above) throughout the test period in both the oxygen and propane injection wells. All oxygen, propane, and first two downgradient transect monitoring wells in the Test Plot received bromide hits. Six of the eight oxygen wells and first two downgradient transect monitoring wells in the Control Plot also received bromide hits. However, all bromide hits occurred at the deep (Blue) screen with one exception: that being an intermediate level screen in the Control Plot. These findings indicate that d-MTBE and tracers in the injection wells were in direct and constant communication with the vendor treatment gases. Furthermore, the tracer injection system delivered the proper concentration consistently and reliably for the duration of the injection period.

Based on the results of the pre-demonstration study, there was sufficient evidence that the system would operate adequately during the long-term monitoring period. Preliminary evaluations of the pre-demonstration bromide tracer study indicated that, aquifer properties in the test areas were consistent with expectations. Therefore, no deviation from the original design of the study was warranted in terms of the rate and volume of samples collected. Furthermore, the pre-demonstration results were used by the SPRD Technical Project Manager (TPM) in designing the monitoring plan for the performance evaluation phase of the project as well as to determine the amount of d-MTBE and iodide necessary for injection. Since it was determined that most ground-water flow occurred in the lower part of the aquifer, the middle and upper wells screens were sampled less frequently. By concentrating on the lower aquifer sampling points, the project was expanded to 15 sampling events rather than the originally planned 7, thereby increasing the statistical strength of the project results.

For the concentrations of the tracer suites, bromide was introduced to the aquifer by adding 352 mg of bromide ion to each well each day. Downgradient bromide concentration values were found to vary significantly, due to the heterogeneity of the aquifer material between sampling points, with no bromide being observed at some locations and concentrations as high as 25 mg/L at transects 2 and 3. Based upon these values, a concentration of 15 mg/L was selected as a representative value for purposes of calculating d-MTBE and iodide injection concentrations. Therefore, the planned d-MTBE and iodide concentrations and additions to the injections wells were based on direct ratios of the bromide results.

The planned concentration of d-MTBE in ground water was approximately 1 mg/L. To obtain this concentration, 23 mg of d-MTBE was introduced into each injection well daily. To assure adequate tracer results, the concentration of iodide was approximately 10 mg/L in the aquifer. To obtain this concentration, 235 mg of the iodide ion was introduced into each injection well daily.

In order to reduce d-MTBE losses to airspace within the tracer reservoir, two 3.8 liter Tedlar bags were used: one for the Control Plot and one for the Test Plot. Consequently, there was no need to use Argon gas as specified in the previous Quality Assurance Project Plan (QAPP). The injection rate of d-MTBE was planned at 10 ml/well/day in order to achieve the desired concentration of approximately 1 mg/L in the aquifer. The total injection volume per plot (Test and Control) for 14-days is equal to 10 ml x 19 wells x 14days which equals 2.66 liters. In order to keep the bags from becoming empty, 3 liters of tracer solution was used as the basis for calculating tracer concentrations in the Tedlar bags used for injection. The d-MTBE was prepared in laboratory ampules for addition to the bags in the field.

Since the injection of oxygen began on May 7, 2001, and the one-time release of the bacterial culture occurred on May 23, 2001, it was assumed that the bacterial culture had been established in the aquifer. Therefore, uniformly labeled d-MTBE and iodide was first introduced to the aquifer on June 8, 2001.

2.3.1 Primary Objective – A Critical Measurement.

The primary project objective was for Envirogen to demonstrate that its in-situ technology could effectively remediate the site under consideration. The effectiveness of the technology was established by examining multiple performance criteria. The critical measurement was whether the levels of d-MTBE and MTBE were significantly less than 5 μg/L in the samples taken over a 10-month period. As shown in

Table 2-4, sampling took place during 15 separate events during the 10-month demonstration, however, only events 4 through 15 were used to evaluate the primary objective.

The primary objective was, therefore, determined by degradation of MTBE in samples collected downgradient of Envirogen's biological barrier over 10 months of continuous operation. This degradation was established by measuring d-MTBE concentrations in the "qualified samples" and determining whether, with 80% confidence, the estimate of the population mean is at or below 5 µg/L. A qualified downgradient monitoring point was one that, at the time of sampling, contained detectable levels of iodide. Iodide and d-MTBE were injected at the onset of the demonstration. During all sampling events, each of the samples collected were evaluated first to assess whether the sample contained iodide at concentrations above the detection limit.

Ground-water samples from all the qualified downgradient monitoring points from the deep screen portion of the aquifer for events 4 through 15 were used for the purpose of quantifying the success of the evaluation. The approved Pre-Quality Assurance Project Plan Agreement (PQA), May 2001, documented the statistical justification and confidence levels associated with the determination of the number of critical samples for the analytes of concern. These samples represented the experimental units for the evaluation of the primary objective.

2.3.2   Secondary Objectives – Non-Critical Measurements

The evaluation was further supported by a number of secondary objectives which provided additional information on treatment processes.

1.   **Determine time of travel to the sampling points using bromide:** The pre-demonstration bromide tracer study was carried out to determine the time of travel to downgradient monitoring wells. This assessment has enabled a more in-depth analysis of ground-water velocity as well as provided information on the hydrologic properties of the system to allow a better understanding of the technology's performance. This information was used to design the sampling plan for the long-term demonstration evaluation.

2.   **Establish the absence of trace metals inhibitors:** During the first sampling event, metals which may inhibit microbial metabolism were assessed to determine potential impact on the technology.

3. **Evaluate the formation of daughter products and determine if they were consistent with a microbiological transformation process:** The biodegradation of MTBE and d-MTBE was expected to result in by-products (i.e., TBA, 2-propanol, acetone, and formaldehyde) some of which may contain deuterium. These products would provide supporting evidence that microbial degradation was resulting from the reduction of MTBE and d-MTBE, as opposed to other abiotic processes (such as dispersion) which could be occurring.

4. **Evaluate changes in geochemical parameters and determine if they were consistent with the microbiological transformation processes:** Parameters such as dissolved organic carbon (DOC), total organic carbon (TOC), alkalinity, sulfide, sulfate, Fe^{++}, conductivity, dissolved oxygen (DO), pH, and water levels were measured to assess whether the ground-water characteristics will be in agreement with the changes expected based on aerobic degradation processes.

5. **Define operating costs over a 10-month period of stable operation:** The cost analysis of the Envirogen technology demonstration at NBVC will be presented along with the unit cost to remediate the ground water contaminated with MTBE per gallon.

6. **Estimate exponential order of degradation and calculate MTBE degradation rate constant:** Using the concentration of d-MTBE from the qualified samples collected during various time intervals, the rate of MTBE degradation would be calculated. A first order degradation of the MTBE concentration will be assumed which allowed the degradation rate to be calculated using the log of the concentration versus time and determining the regression coefficient using a least squares analysis.

7. **Determine the fraction of d-MTBE removed at each sampling location at each sample time:** Using the concentration of d-MTBE in samples collected during the 15 sampling events (Table 2-4), the change in the ratio of d-MTBE concentrations to the tracer concentrations were determined, in accordance with the following equation:

    $$[(d\text{-}MTBE_i/halide_i)/(d\text{-}MTBE_s/halide_s)]$$
    where "i" is the initial concentrations of the tracers and "s" is the
    subsequent values

    Samples found to meet the qualifications for MTBE reduction due to degradation (i.e., samples containing iodide) will be used in determining the degradation rate constant. The purpose of this objective is to compare the relative magnitude of biodegradation between sampling events. Consider the following examples:

    A. At time (i) the d-MTBE concentration at a specific sampling point is 1 ppm and the halide concentration is 10 ppm, and that at time (s) the d-MTBE concentration is also 1 ppm and the halide concentration is 10 ppm. This would yield a ratio of 1 and indicate that no degradation has occurred.

B. At time (i) the d-MTBE concentration at a specific sampling point is 1 ppm and the halide concentration is 10 ppm, and that at time (s) the d-MTBE concentration is 0.5 ppm and the halide concentration is 5 ppm. This would also yield a ratio of 1 and indicate that no biodegradation has taken place even though there was a reduction in d-MTBE from 1 ppm to 0.5 ppm. A similar relative reduction in the halide tracer indicates that a portion of the ground water was diverted from the sampling point thus resulting in lower concentrations of d-MTBE.

C. At time (i) the d-MTBE concentration at a specific sampling point is 0.5 ppm and the halide concentration is 5 ppm, and at time (s) the d-MTBE concentration is 0.1 ppm and the halide concentration is 5 ppm. This would yield a ratio of 5 and indicate that there is a relative 5x degradation of d-MTBE.

D. At time (i) the d-MTBE concentration at a specific sampling point is 0.1 ppm and the halide concentration is 5 ppm, and at time (s) the d-MTBE concentration is 0.5 ppm and the halide concentration is 5 ppm. This would yield a ratio of 0.2 which also indicates a 5x change; however, the level of degradation has decreased.

Therefore, the equation will be used to calculate the ratio between sampling events at each location and interpreted as follows: A ratio of 1 indicates no relative change in degradation. Ratios of >1 indicate a relative increase in degradation between events. Ratios of <1 indicate a relative decrease in degradation between events. These ratios will be mapped to visualize the relative spatial changes in degradation.

8. **Evaluate d-MTBE reduction in the Control Plot receiving only oxygen injection.** Tracers were injected into the Control Plot in the same manner as the Test Plot allowing an assessment of changes in MTBE and d-MTBE concentrations as a result of indigenous bacteria relative to the changes observed in the Test Plot by the exogenous microflora. According to the PQA, concentrations of d-MTBE between the Test Plot and Control Plot will be examined statistically.

Table 2-1
Summary of Site Characterization Analytical
Results For Contaminants of Concern at The Middle Zone

Pre-characterization Sampling Event, September 2000

Well Number	Sample ID	MTBE µg/L	TBA µg/L	Ethyl benzene µg/L	Benzene µg/L	Toluene µg/L	Total Xylenes µg/L
1	MW - 1	3,750	11.1	ND	ND	ND	ND
2	MW – 2	4,650	12.3	ND	ND	ND	ND
3	MW – 3	4,090	13.7	ND	ND	ND	ND
4	MW – 4	591	ND (10)	ND	ND	ND	ND
5	MW – 5	296	ND (10)	ND	ND	ND	ND
6	MW – 6	341	ND (10)	ND	ND	ND	ND
7	MW – 7	5,040	17.4	ND	ND	ND	ND
8	MW – 8	3,250	ND (10)	ND	ND	ND	ND
9	MW – 9	3,260	10.5	ND	ND	ND	ND
10	MW - 10	1,900	ND (10)	ND	ND	ND	ND

Results from Method Validation Study, November, 2000

Laboratory Number	Sample ID	MTBE µg/L	TBA	2-Propanol µg/L	Acetone µg/L	BTEX µg/L
Lab 1	GWT (lab 1)	2960 - 3010	ND – 12	ND	ND	NA
Lab 2	GWT (lab 2)	2500 -2900	ND – 69	ND	ND	NA
Lab 3	GWT (lab 3)	2200 -2840	60 –80	ND	ND	NA
Lab 4	GWT (lab 4)	2200 - 2550	10 – 21	ND	ND	ND

µg/L   micrograms/Liter
NA   not analyzed
ND   not detect

Table 2-2

Analyses to Support the Propane Biostimulation and Bioaugmentation Project Objectives

Matrix	Parameter	Classification	Type	Purpose
Ground Water (upgradient and downgradient monitoring wells in treatment and control plots)	Deuterated-MTBE/MTBE	Critical	Analytical	1° objective: MTBE reduction.
	Iodide	Critical	Analytical	1° objective: MTBE reduction. 2° objective: ground-water flow field determination
	Deuterated-TBA/TBA	Non-critical	Analytical	2° objective: Evaluate biotic generation of MTBE daughter products
	Deuterated-Acetone/Acetone	Non-critical	Analytical	
	Deuterated-2-propanol/2-propanol	Non-critical	Analytical	
	Formaldehyde	Non-critical	Analytical	
	Alkalinity	Non-critical	Analytical	2° objective: Geochemical indicators of ground water .
	TOC	Non-critical	Analytical	
	DOC	Non-critical	Analytical	
	Conductivity	Non-critical	Field	
	Temperature	Non-Critical	Field	
	Sulfide	Non-critical	Field	
	PH	Non-critical	Field	
	Fe(II)	Non-critical	Field	
	Water Level	Non-Critical	Field	
	Dissolved Oxygen	Non-critical	Field	2° objective: determine presence of treatment gas in ground water.
	Phosphate	Non-critical	Analytical	2° objective: Evaluate characteristics for support of biological processes
	Sulfate	Non-critical	Analytical	
	Nitrate	Non-critical	Analytical	
	Nitrite	Non-critical	Analytical	
	Metals	Non-critical	Analytical	
	Ammonia	Non-critical	Analytical	

Note:

1   Biostimulation included application of oxygen and propane to the Test Plot, and oxygen to the Control Plot.

2   Bioaugmentation included one time release of Envirogen propane oxidizing bacteria (POB) *Rhodococcus ruber* strain ENV425 into the aquifer at the Test Plot. The Control Plot was devoid of exogenous bacteria.

Table 2-3
Applicable Regulatory Criteria for
MTBE Treatment Technology Demonstration Program

PARAMETER GROUP		CA Primary MCL MCL (mg/L)	CA Secondary MCL (mg/L)	CA Action Level (mg/L)	CA Public Health Goal (mg/L)	Stage 2 DBPR MCLb (mg/L)	Demonstration Treatment Goal (mg/L)
Volatile organics							
	MTBE*	0.013	0.005	NA	0.013	NA	0.005
	TBA	NA	NA	0.012	NA	NA	0.012
	Acetone	NA	NA	NA	NA	NA	NA
	Benzene	0.001	NA	NA	0.00014^b	NA	0.001
	Toluene	0.15	NA	NA	0.15	NA	0.15
	Ethylbenzene	0.7	NA	NA	0.3	NA	0.7
	Xylene^c	1.75	NA	NA	1.8	NA	1.75
DW parameters (SDS Testing)							
	TTHMS	0.1	NA	NA	NA	0.08	0.08
	HAAs	NA	NA	NA	NA	0.06	0.06
	NDMA	NA	NA	0.00002	NA	NA	0.00002
Aldehydes/glyoxals							
	Formaldehyde	NA	NA	NA	NA	NA	NA
	Acetaldehyde	NA	NA	NA	NA	NA	NA
	Heptaldehyde	NA	NA	NA	NA	NA	NA
	Glyoxal	NA	NA	NA	NA	NA	NA
	Methyl glyoxal	NA	NA	NA	NA	NA	NA
Wet chemistry							
	Bromate	NA	NA	NA	NA	0.01	0.01

Abbreviations:

CA: State of California
DBPR: Disinfection Byproduct Rule
DO: Dissolved oxygen
DW: Drinking water

HAAS: Haloacetic aids
MTBE: Methyl *tert*-butyl ether
NA: Not Available

SDS: Simulated Distribution System
TBA: *tert*-Butyl alcohol
TTHMs: Total trihalomethanes

Notes:
* Critical parameter (associated with a primary Envirogen demonstration objective)

a.) Sources: California DHS:

Stage 2 DBPR

Primary MCLs and Lead and Copper Action Levels (January 2001), Secondary MCLs (May 2000), Action Levels (February 2001), Public Health Goals (January 2001)

Stage 2 Microbial and Disinfection Byproducts Federal Advisory Committee Agreement in Principle; 65 FR 83015 (December 29, 2000)

b.) Draft or proposed values.
c.) Single isomer or sum of isomers

2.4   SCHEDULE

The demonstration of Envirogen oxygen and propane biostimulation and bioaugmentation technology was implemented in one phase; the technology was evaluated over a ten-month period during June 2001 – March 2002 as shown in Table 2-4. While the injection of the vendor's gaseous substrates (oxygen and propane), and SPRD tracers (d-MTBE and iodide) starting dates are provided in Table 2-4, their release continued throughout the project period.

Prior to the start of the demonstration, Envirogen conducted a background-sampling event on May 2, 2001, and requested a two-week period for the sparging system optimization (May 7 – May 21). At the conclusion of this period, a second background-sampling event on May 22, was followed by the injection of an exogenous culture on May 23, 2001. After the aquifer was presumably equilibrated, the injection of EPA's tracer started on June 8, 2001. Envirogen and EPA's first sampling event were conducted on June 12, 2001, and June 14, 2001, in accordance with the time of travel established during the bromide tracer study.

Table 2-4
U.S. EPA Performance Monitoring
Sampling Schedule

Event	Description	Date
Process Optimization	Oxygen Sparge	05/07-20/01
Treatment	Propane and Oxygen Sparge	05/21/01
Treatment	Bioaugmentation*	05/23/01
Tracer	d-MTBE/Iodide	06/09/01
Event 1	GW Sampling	6/14/01
Event 2	GW Sampling	06/28/01
Event 3	GW Sampling	07/09/01
Event 4	GW Sampling	07/17/01
Event 5	GW Sampling	07/30/01
Event 6	GW Sampling	08/13/01
Event 7	GW Sampling	08/27/01
Event 8	GW Sampling	09/10/01
Event 9	GW Sampling	09/24/01
Event 10	GW Sampling	10/08/01
Event 11	GW Sampling	11/05/01
Event 12	GW Sampling	12/03/01
Event 13	GW Sampling	01/07/02
Event 14	GW Sampling	02/11/02
Event 15	GW Sampling	03/08/02

GW: Ground Water
*: Bioaugmentation included a one-time release of Envirogen propane oxidizing bacteria (POB)
Rhodococcus ruber strain ENV425 into the aquifer at the Test Plot. The Control Plot was
devoid of exogenous bacteria.

SECTION 3

PERFORMANCE MONITORING APPROACH

The demonstration objectives outlined previously in Section 2 were achieved by a carefully designed performance monitoring program, developed by the SPRD TPM, which provided data to evaluate the technology. The evaluation of the Envirogen technology was accomplished by the sampling and analysis of both the Test and Control Plots illustrated in Figure 3-1. It is important to note that throughout the report the well-screen designations are: Red (R) for shallow; Yellow (Y) for middle; and Blue (B) for deep locations. Both plots included a series of well clusters to allow the collection of ground-water samples at discreet depth intervals. Sample collection and analysis, in support of the various objectives, included a suit of tracers injected as an integral part of this evaluation. The following sections provide details on the experimental design used in the evaluation of Envirogen's biostimulation technology. These sections include a discussion of how the design achieved project objectives, a description of the layout of the Test and Control Plots and a description of the tracer injection system.

3.1   TRACER STUDY DESIGN COMPONENETS

The Envirogen technology demonstration was designed to determine the efficacy of using propane and oxygen biostimulation and bioaugmentation (exogenous propane oxidizing bacteria *Rhodococcus ruber* strain ENV425) as a potential remedial alternative for the removal of MTBE from ground water. Achieving this objective resulted, in great measure, from the use of d-MTBE and ground-water tracers. The ratios of ground-water tracers between downgradient transects provided evidence concerning the relative losses in MTBE concentrations resulting from dilution and degradation. Likewise, the use of d-MTBE ratios in downgradient transects served as a tracer of anthropogenic MTBE. More importantly, the use of d-MTBE provided evidence of biodegradation by the realization of d-MTBE daughter products.

A critical element of this evaluation was the utilization of a ground-water tracer mixing and injection system. It was essential for delivering mixed tracers into the ground water at both the Test and Control Plots. A uniformly labeled d-MTBE was introduced through the ground-water tracer mixing and injection system, upgradient of the oxygen and propane injection location at a concentration of approximately 1 mg/L. Section 2.3 discusses specifics of the tracer injection system that was designed to be operational for the duration of the demonstration.

30

The ground-water tracer mixing and injection system consisted of 19 well injection points directly upgradient of the oxygen and propane injection points in both the Test and Control Plots (total of 38 wells). Tracer materials, including d-MTBE and iodide or d-MTBE, iodide, and bromide were metered into each well so as to deliver a known and constant concentration into the ground water. A pre-demonstration ground-water flow study, using bromide as the tracer, was implemented prior to the start-up of the vendor's technology to document natural ground-water flow gradients and provide a baseline for the tracers in downgradient monitoring points. A configuration of the tracer mixing and injection system is shown in Figure 3-2. Figure 3-3 depicts the cross section of the tracer circulation components. Each injection well was installed to a depth of 24 feet BGS. The tracers were introduced to each well by a metering pump connected to a 1/8" stainless steel line which discharged to the well below the top of the water table. Seasonal ground-water fluctuations were considered prior to positioning the line. In order to recirculate ground water at approximately 100 milliliters per minute, each well was equipped with a bladder pump. Tubing was used to connect the bladder pumps to the compressor via air controllers. Each controller operated two pumps and a total of 20 controllers were employed for the construction of the system. Two 40-channel metering pumps were employed for the injection of the tracers into the thirty-eight 2-inch wells at an approximate flow rate of 10 ml/well/day.

A multiple-tracer approach was employed (Thierrin et al., 1992, 1993, and 1995; Poulson et al., 1997; Aeschbach-Hertig et al., 1998; Davis et al., 2000; Parker and van Genuchten, 1984; Kenoyer, 1988; Melville et al., 1991; Meiri, 1989; Bowman and Gibbens, 1992; Bullivant and O'Sullivan, 1989; Stute et al., 1987; Patrick and Barker, 1985; Gupta et al., 1994; Poulson et al., 1995; Stevenson et al., 1989). Iodide or iodide and bromide were used in response to abiotic issues and deuterated-MTBE was used to reliably quantify the biotic challenges. The d-MTBE tracer was identical to the primary contaminant of concern. Since in practice, deuterated and non-deuterated species have the same fate and solute transport properties (i.e., sorption, desorption, biodegradation), they can be employed like internal standards for the assessment of in-situ intrinsic or enhanced bioremediation. The use of d-MTBE provided evidence to determine if microbes metabolize MTBE partially to TBA (incomplete biodegradation) or completely to CO_2. The d-MTBE migrated with the ground water containing the intrinsic MTBE. It was expected that the cometabolic degradation (Garnier et al., 1999; Hyman et al., 1998) of d-MTBE would result in daughter products (e.g., TBA, 2-propanol, acetone) containing deuterium which can easily be determined by gas chromatography/mass spectroscopy (AWWA, 1998; Bianchi and Varney, 1989; Bonin et al., 1995; Church et al., 1997a,b; Kanal et al., 1994; Nouri et al., 1996;). The use of d-MTBE was accepted by the stakeholders as a way to assure that a reduction in MTBE can be demonstrated to be a result of microbial

31

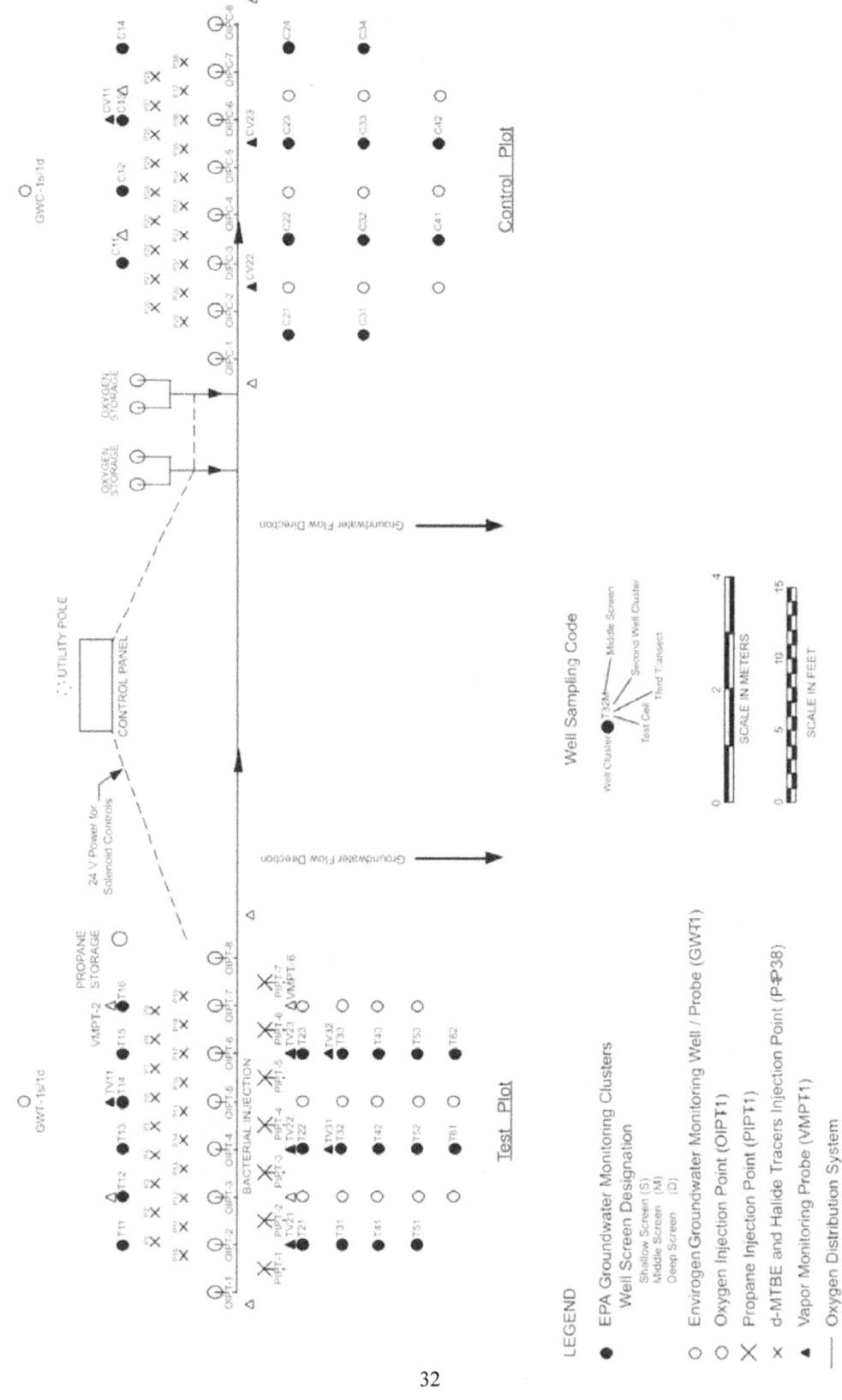

Figure 3-1. Test and Control Plots Layout

(Reprinted from *Performance Monitoring of Enhanced In-Situ Bioremediation of MTBE in Ground Water*. Draft Work Plan. 2000, by Ann Keeley, SPRD, USEPA.)

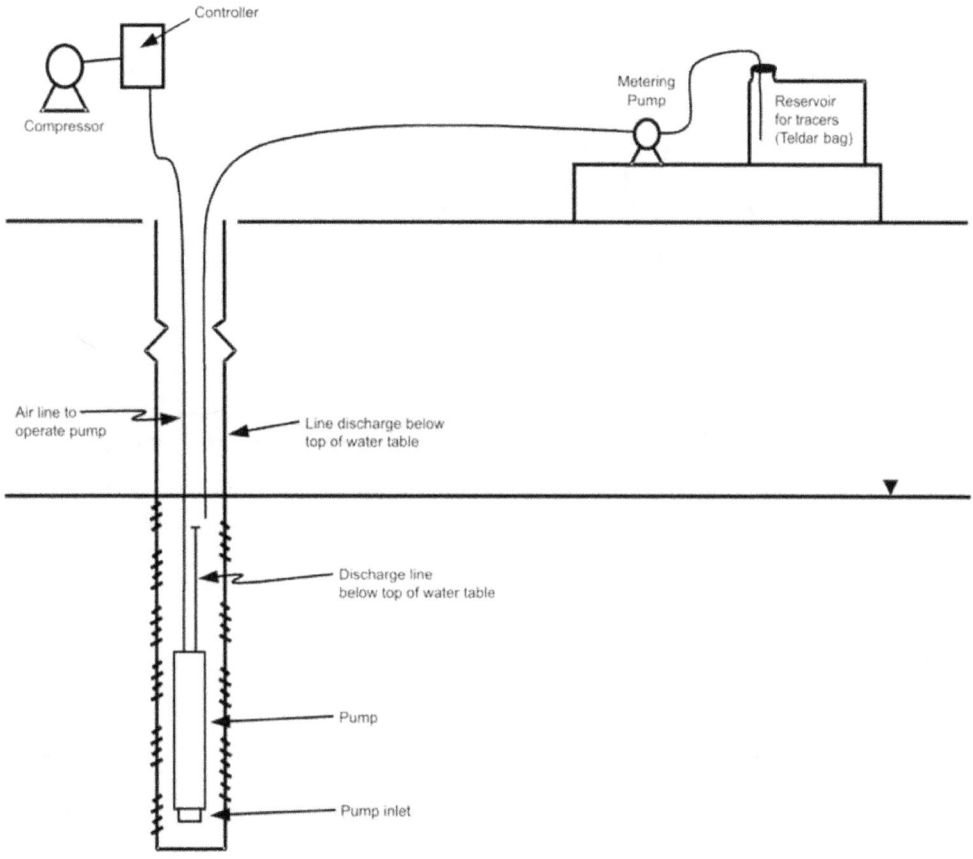

Figure 3-2. Tracer Circulation Well

(Modified from *Performance Monitoring of Enhanced In-Situ Bioremediation of MTBE in Ground Water*, Draft Work Plan, 2000, by Ann Keeley, SPRD, USEPA.)

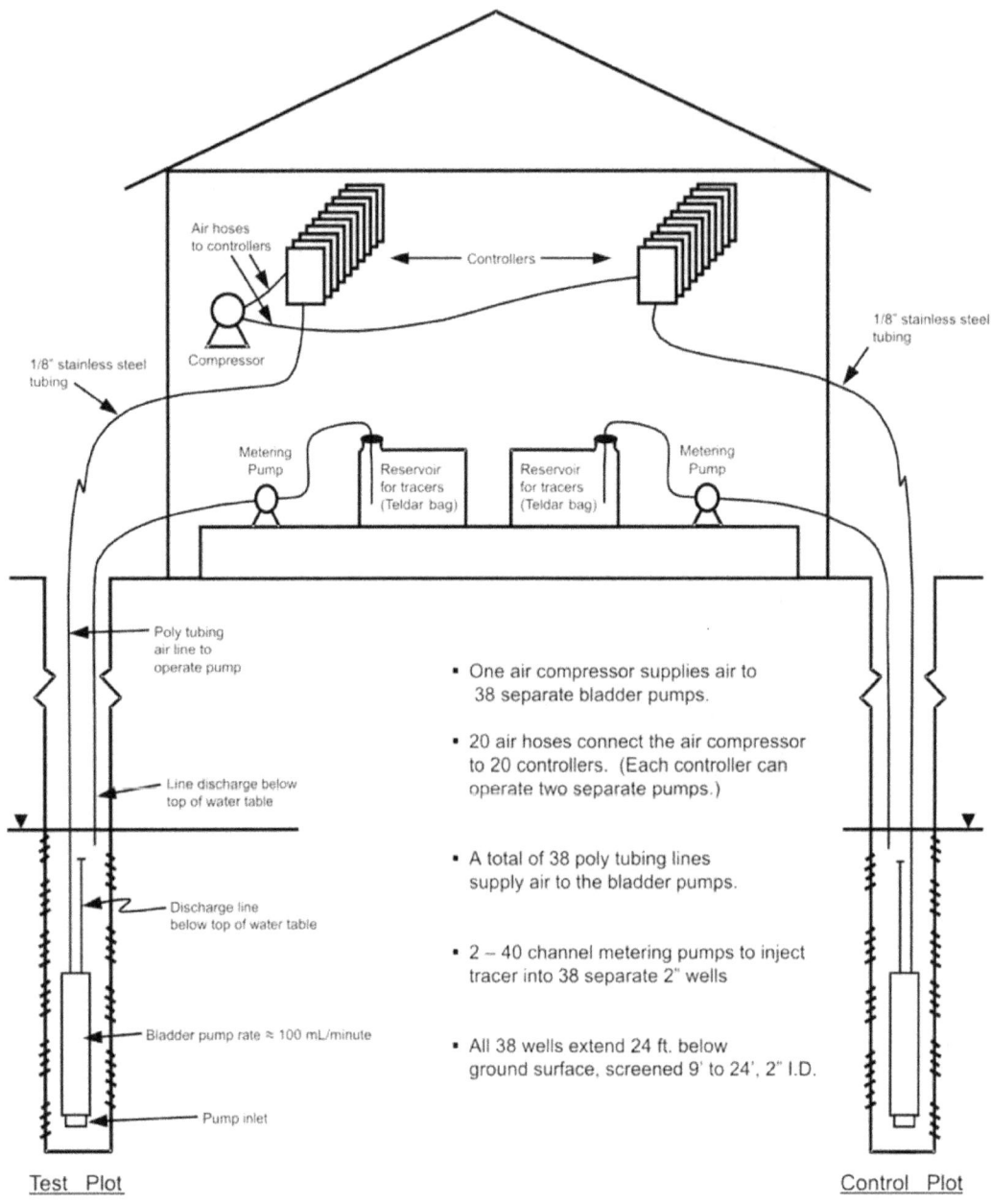

Figure 3-3. Tracer Circulation System Cross Section

(Modified from *Performance Monitoring of Enhanced In-Situ Bioremediation of MTBE in Ground Water*, Draft Work Plan, 2000, by Ann Keeley, SPRD, USEPA.)

degradation rather than other processes such as dilution (since biosparging activity could be displacing the MTBE contaminated ground water vertically and/or laterally) (Borden et al., 1997; Bernauer et al., 1998; Church et al., 1999a,b, and 2000; Hanson et al., 1999; Hardison et al., 1997; Jensen and Arvin, 1990; Mo et al., 1999; Connell, 1994; Ronen et al., 1993; Schirmer et al., 1999; Suflita and Mormile, 1993; Yen and Novak, 1994; White et al., 1996). In addition, the tracer injection system continuously delivered d-MTBE at approximately 1 mg/L to the biological barrier as a challenge material for the technology evaluation.

Ground-water tracers were used to establish the direction and velocity of ground-water movement. Two halides, which were present at background quantities (i.e., bromide and iodide), were added at upgradient locations using the 38 tracer circulation wells: bromide prior to the start of the treatment and the iodide, or iodide and bromide after the on-set of the treatment. As previously described in Section 2.3, the approximate concentrations of iodide and d-MTBE were 10 mg/L and 1 mg/L, respectively.

3.1.1   Test and Control Plot Design

The Test and Control Plots were placed within the Middle Zone, a portion of the aquifer impacted by moderate levels of MTBE with no BTEX compounds. The placement of the plots enabled an assessment of the technology as a biological barrier to the migration of MTBE from the upgradient source area. Figure 3-1 depicted the configuration of each plot within the study area, the type and position of monitoring points, the position of the vendor's injection system, and the placement of the tracer injection system. The plots (monitoring network transects) were aligned perpendicular to the ground-water flow. According to the site pre-characterization activities, ground-water gradients were determined to be in the direction shown in Figure 3-1. Each plot measures 30 feet wide by 40 feet long and they are separated by approximately forty feet. Both the Test Plot and Control Plot contained upgradient and downgradient transect well clusters (Smith et al., 1991). For the evaluation study, only SPRD monitoring points were sampled. Envirogen collected samples from their designed monitoring points.

For the Test Plot, there were a total of six upgradient ground-water monitoring clusters located in a single transect. Each cluster contained one well screened near the bottom of the aquifer (21" screen), one screened in the middle of the saturated zone (6" screen), and one screened near the top of the water table (21" screen). Downgradient of the biosparging and tracer injection systems, there were a total of 14 ground-water monitoring clusters located along five parallel transects. Each cluster contained wells screened near the bottom, middle, and the top of the aquifer. Each screen was 27". The exact position of the screens was determined based on vertical hydraulic conductivity profiling prior to installation (See Figure 3-4, Well Construction Specifications).

For the Control Plot, there were a total of four upgradient ground-water monitoring clusters located in a single transect and ten downgradient monitoring clusters located along three parallel transects. Transects for the Control Plot did not necessarily align with similar transect numbers for the Test Plot. The well screen locations for both the upgradient and downgradient were located as described for the Test Plot.

3.1.2   Monitoring Parameters

Although various parameters were measured during this performance validation study, the determination of the MTBE/d-MTBE and iodide were more significant in terms of their role as the critical measurements in support of the project's primary claims. Nevertheless, all other listed analytes (i.e., TBA, acetone, 2-propanol, and their deuterated forms) were measured and were used collectively as supportive evidence for the biodegradation potentials. All samples were collected as described in Section 4 and analyzed according to the procedures referenced in Table 4-1.

3.1.3   Sampling Approach

There were a total of fifteen sampling and analysis events. The first three took place in the first month after the addition of tracers, as described in the schedule in Section 2.3. The final sampling took place at approximately 10 months (event 15) after the initiation of treatment (Table 2-4 details the schedule for all the other intermediate events).

Selected wells from both the Control and Test Plots were sampled during each event (see PQA NRMRL QA Number 119-Q12 and QAPP Appendix B for specifics). Various parameters were sampled and analyzed during each event, as noted in subsequent sections of this document. Each of the screened depth intervals in the well clusters were considered to be independent for this demonstration and were based on the radius of influence during sampling which was a function of the ground-water velocity.

Based on the results of the November 2000 aquifer tests conducted by SPRD hydrogeologists within the demonstration plot, the sampling plan called for the completion of an event in three days or less using a pumping rate of 30 ml/min.

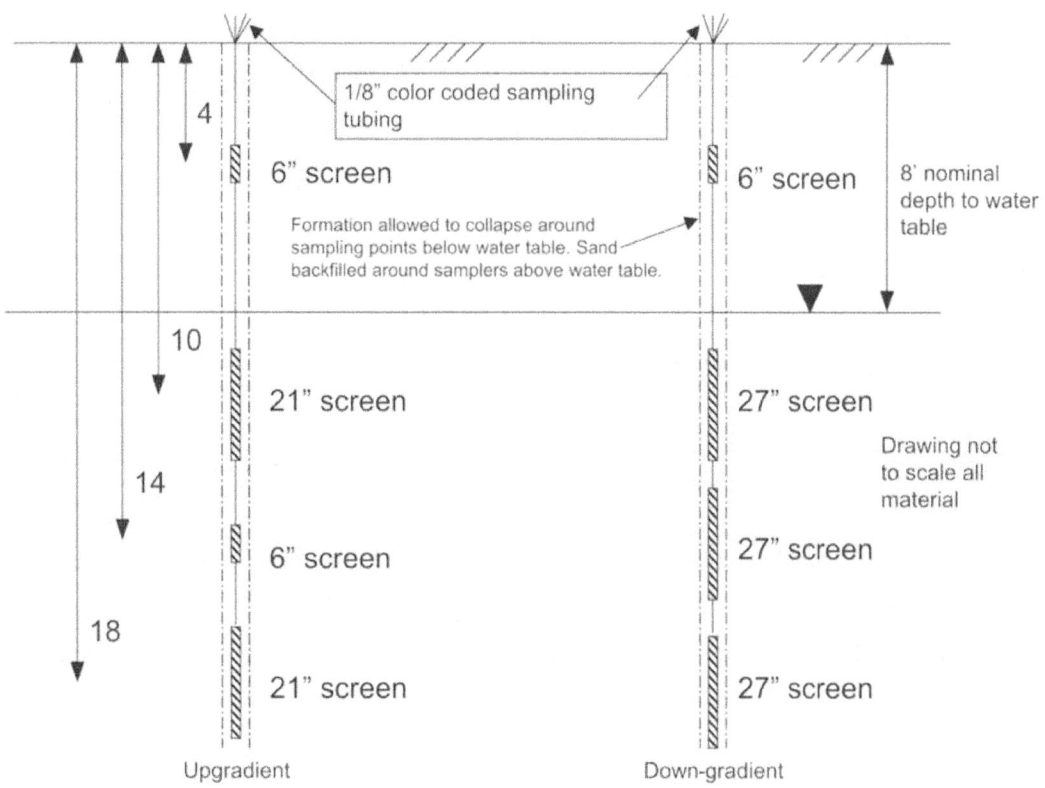

Figure 3-4. Well Construction Specifications

(Reprinted from *Performance Monitoring of Enhanced In-Situ Bioremediation of MTBE in Ground Water*, Draft Work Plan, 2000, by Ann Keeley, SPRD, USEPA.)

SECTION 4
SAMPLING AND ANALYSIS PROTOCOL

The following section provides details on the procedures used to collect samples for the SPRD long-term performance monitoring of the Envirogen demonstration. The collection of ground-water samples for d-MTBE and iodide was critical for the achievement of the project's technical objectives.

4.1   GROUND-WATER SAMPLING

As noted earlier (Section 3 and Figure 3-1), there are a series of ground-water monitoring well clusters in the Test Plot as well as the Control Plot, both upgradient and downgradient of the oxygen (Control Plot) and oxygen/propane and bacterial injection wells (Test Plot). In the Test Plot there are 6 upgradient wells and 14 downgradient SPRD wells. In the Control Plot there were 4 upgradient wells and 10 downgradient SPRD wells. These wells, constructed as detailed below, were sampled during 15 events for a variety of parameters in accordance with the schedules provided in the following subsections.

4.1.1   Monitoring Well Specifications

The installation of monitoring wells and tracer injection wells was accomplished using a Geoprobe™ unit and Cone Penetrometer Technology (CPT). The upgradient wells were installed using a Geoprobe™ rig while the downgradient wells were placed with the CPT. Initially, 3.25" O.D. or 2.125" O.D. rods with an expandable point were advanced to a predetermined depth. The monitoring well was assembled then lowered inside the probe rods which were then retracted. As the rods were retracted, the natural formation was allowed to collapse around the well. Sand was used to backfill around the wells above the water table with a six-inch bentonite plug placed at the surface. After installation, the wells were developed by purging and mechanical surging following ASTM D 5521 "Standard Guide for Development of Ground-Water Monitoring Wells in Granular Aquifers."

4.1.2   Low Flow Sampling

The primary limitations to the collection of representative ground-water samples include: disturbance of the water column above the screened interval; re-suspension of settled solids at the base of the casing (e.g., high pumping rates); disturbance at the well screen during purging and sampling (e.g., high pump rates); and the introduction of atmospheric gases or degassing from the water (e.g., sample handling, transfer, vacuum from sampling device, etc.). Based on these limitations, low-flow sampling protocols were employed for this demonstration evaluation.

4.1.3   Well Purging

The purging of monitoring wells for the purpose of obtaining representative samples is necessary since ground-water chemistry can be altered through contact with the atmosphere, well casing materials, screen, and surface seal. However, due to flow limitations, sample volume restrictions and the desire to obtain samples with as little disturbance as possible, a flow-through cell to establish well stabilization was not used. The purge volume for each well was 30 mls. Low flow rates were used, both during purging and sampling.

To achieve minimum disturbance to the formation, direct push technology was used to install the monitoring points. Since the wells were completed using natural formation collapse, the screen was in direct contact with formation water. Based upon the experiences of SPRD personnel at a similar site using identical screen types, it was determined that pumping one pore volume would provide a representative sample of the intrinsic ground water. A peristaltic pump was used for purging and sample collection. After 30 ml had been purged, sampling was initiated. The purge volume was based on purging 1.5 times the well bore volume of the screened interval and the associated tubing.

4.1.4   Well Sampling

The ground-water monitoring wells were 1/8-inch diameter stainless steel with tubing connected at the top of each for sampling. Based on the results of the November 2000 aquifer tests conducted by SPRD hydrologists within the demonstration plot, the sampling plan called for the completion of a full sampling event in three days or less using a pumping rate of 30 ml/min. Therefore, approximately one liter of water was collected from each well per sampling event at a flow-rate of 30 ml/min.

Due to concerns associated with the loss of VOCs, after the third sampling event, samples were collected at 30, 50, and 90 ml/min to determine an appropriate sampling rate. It was determined that a sample rate of 50 ml/min for the shallow and intermediate zones and 90 ml/min for the deep zone would be used for the remainder of the demonstration.

In order to confine each sampling event to three days or less, sampling began at the line of wells farthest downgradient from the injection wells then moved to the next upgradient line of wells. Samples were taken first from the deepest ground-water horizon until all of the wells had been sampled at that location, followed by the middle horizon, and finally the top horizon. This procedure minimized disturbances to the flow field and negates artifacts imposed by the sampling process. A trip blank accompanied each shipment of VOC samples to the laboratory.

Table 4-1 indicates the parameters and analytical methods used for determination of the listed analytes. Due to the limitations in the amount of water that could be collected the bottles were filled in the following order:

1. Ammonia
2. Alkalinity
3. Dissolved gases
4. VOCs (MTBE, TBA, 2-propanol, acetone and deuterated isotopes)
5. Iodide
6. Formaldehyde
7. TOC/DOC
8. Nutrients
9. Metals

When samples for specific analytes were not scheduled to be collected, the sample stream was delivered to a waste container for that analyte and then redirected back to the next sample bottle after the appropriate volume of water had passed. Each water sample for VOC analysis was collected in two 40 milliliter volatile organic analysis (VOA) vials containing hydrochloric acid to acidify the sample to a pH of less than 2. The water sample was gently introduced into the sample containers to reduce agitation and loss of volatile compounds. Each vial was filled until a meniscus appears over the top of the vial. The screw-top lid with the septum (Teflon side toward the sample) was then tightened onto the vial. After the lid was tightened, the vial was inverted and tapped to check for air bubbles. If any air bubbles were present, the sample was recollected. For all other analytes, water was introduced directly into the appropriate container, as listed in Table 4-1, and the lid was tightened immediately after filling. Field duplicates and other quality control (QC) samples were collected immediately following collection of the original sample. After collection, each water sample was stored on ice in a cooler until readied for overnight shipment to the analytical laboratory. An exception was the formaldehyde samples that were immediately being picked up by a DelMar laboratory dispatcher. All sample collection procedures were in accordance with the reference method listed in Table 4-1. Following sample collection, each sample was labeled with detailed information regarding the location, date, and time of collection. Chain-of-custody procedures were followed from sample collection through sample analysis.

In addition to the designed analytes that were used for the evaluation of the technology's performance within the context of this ITER, shortly after the completion of each EPA sampling event, the Navy personnel conducted the following ground-water quality field measurements: conductivity, sulfide, pH, Fe^{++}, dissolved oxygen (DO), water table elevations, and temperature from the selected Envirogen monitoring wells. The determination of these measurements involved purging for stability to provide an insight as to the general quality of the aquifer.

4.2   TRACER INJECTION SYSTEM OPERATIONS AND MAINTENANCE

As discussed in Section 3.1, the tracer injection system played an integral role in the operation of the demonstration assessment investigation. The 19 wells in both the Control and Test Plots were used in the initial bromide tracer study to characterize ground-water flow paths and monitor tracer concentrations. During the evaluation phase of the project, the wells were used to inject the non-conservative and conservative tracers d-MTBE and iodide.

In order to establish that the proper function of the tracer injection and circulation system was maintained during the entire evaluation process, SPRD and NBVC personnel performed routine maintenance work prior to injecting the d-MTBE/iodide tracer solution as well as during and after the termination of the project. In addition to close-of-day observations of the tracer reservoirs and metering pumps, a regimented operation and maintenance program included; multiple flow rate tests at the wellhead to insure the proper function of the metering pumps in delivering an exact amount of tracer solution; monthly replacement of the Tygon tubing and an inspection of each of the balder pumps; monthly ground-water sampling for d-MTBE/MTBE, iodide; and measurement of circulation flow rate prior to and after the termination of the project within the 38 injection wells. Water levels and geochemical parameters were established prior to the start of the demonstration and at its conclusion. Periodically performed tests included the determination of water levels and dissolved oxygen in the injection wells. A brief description of system maintenance, with an emphasis on activities prior to the system startup, is provided below.

Prior to the long-term performance monitoring and tracer system start up, on May 30, 2001, ground-water samples were collected to establish intrinsic MTBE and bromide background. Furthermore, to confirm that the previously determined flow rate of 10 ml/well/day was being injected, a number of bench and field-scale flow rate tests were conducted from June 4 – 8, 2001. During these tests, the collected reservoir solution volumes were measured and the measurements were used to determine or adjust the pump settings. The field testing flow rate was confirmed by setting the metering pumps on 2.9 – 3.0 revolution per minute (rpm).

On June 8, 2001, the pumps were shut off and downhole tubing was disconnected. After all lines to the wellhead were filled with the tracer solution, two samples were collected, each from the Test and Control Plots for d-MTBE and iodide analysis. After the system adjustment, a 12 hours flow rate test was conducted. During the test, one end of each tracer delivery line was inserted into a 10 ml VOA vial at the wellhead, while the other end was connected to the metering pumps. Each VAO vial was then secured to the wellhead. The flow rate at the wellhead was determined based on the volume of tracer solution accumulated (total of 38 vials). The proper final readjustments to the pump settings were made. The above procedure was repeated during a 4-hour rate measurement that resulted in the replacement of some Tygon tubing. Based on the result of the multiple flow rate tests, final metering pump settings of 2.9 – 3.0 rpm was revalidated based on the volume of the tracer collected in the VOA vials. The tracer reservoirs were then sampled for d-MTBE and iodide just prior to the start of the injection process.

On June 9, 2001, the SPRD long-term injection of tracer solution containing d-MTBE and iodide was initiated and continued during the entire demonstration project. During a follow-up system inspection on June 10, 2001, the visual examination of the Test and Control Plot metering pump components concluded that the system was not clogged. The inspector recorded that the system was working properly after pulling out and inspecting all of the tubing. The inspection of the submersible pumps and control also were recorded as working properly.

At the onset of the EPA critical sampling, Event 4, a 5-hour flow test was also conducted on July 16, 2001, to determine if the optimal tracer delivery rate was maintained. If not, the sampling event was to be terminated. This test, which superseded the QAPP sampling, demonstrated that the precise volume of the tracer solution was being pumped into every injection well. The volume collected from each of the wells was 2 ml ($\pm$ 100 μl) which is consistent with the projected volume. Therefore, the metering pumps were operating correctly.

As stipulated by the field Technical System Audit (TSA), the flow rate of the tracer injection pump was measured at the termination of the Envirogen demonstration. In an attempt to document the rate of the tracer injection at the wellhead, on March 12, 2002, after the completion of the QAPP sampling, a 12-hour rate test was conducted. It is noted that during this period, the tracer injection into the well bores was interrupted, as discussed in Section 6.2.2. As a result, 5 ml of tracer was collected at each well head ($\pm$ 100 - 200 μl). This indicated that the flow rate was producing the required 10 ml/day.

Table 4-1
Analytical Parameters and Method Requirements

Target Analytes	Type (4)	Method	Minimum Volume	Preservation[a]	Holding Time	Analytical Laboratory
MTBE / d-MTBE (1)	C	SW846-5030/8260B	2 x 40 mls	HCL to pH<2	14 Days	ALSI
TBA/d-TBA (1)	NC	SW846-5030/8260B				ALSI
2-propanol/ d-2-propanol	NC	SW846-5030/8260B				ALSI
Acetone/ d-acetone	NC	SW846-5030/8260B				ALSI
Nitrate/Nitrite	NC	EPA 300.0	2 x 40 mls	None	2 Days	ManTech
Sulfate	NC	EPA 300.0			28 days (same sample as above)	ManTech
Phosphate	NC	EPA 300.0			2 Days (same as above)	ManTech
Ammonia	NC	EPA 350.3	50 mls	H_2SO_4 to pH<2	28 Days	ALSI
Alkalinity	NC	SM 2320B	100 mls	None	14 Days	ALSI
TOC	C	SW 846/9060	2 x 40 mls	HCL to pH<2	28 Days	ALSI
DOC	NC	SW 846/9060	2 x 40 mls	None (6)	28 Days	ALSI
Formaldehyde (2)	NC	SW 846-8315	200 mls	None	3 Days (ext.)(2)	DMA
Iodide/Iodate (3)	NC	EPA 300.0	2 x 40 mls	None	7 Days (3)	ManTech
Metals (5)	NC	SW 846 3010/6010	100 mls	HNO_3 to pH<2	6 Months	ALSI
Mercury	NC	CVAA 7470	100 mls	HNO_3 to pH<2	28 Days	ALSI

Abbreviations:

ALSI: Analytical Services, Inc.
DMA: DelMar Analytical Services
DOC: Dissolved organic carbon
d-MTBE: Deuterated methyl *tert*-butyl ether
d-TBA: Deuterated *tert*-butyl alcohol
HCL: Hydrochloric acid

HNO_3: Nitric acid
mls: milliliters
MTBE: Methyl *tert*-butyl ether
SW 846: Test Method for the Evaluation of Solid Wastes (EPA 1996)
TBA: *tert*-Butyl alcohol
TOC: Total organic carbon

Notes:

a) In addition to the chemical preservation methods indicated above, all samples were cooled to 4^0C for shipment and storage.

(1) Includes deuterated forms MTBE-d_3 and TBA-d_{10}, d-MTBE is only critical

(2) Liquid-liquid (separatory funnel) was used for the extraction. Formaldehyde hold time is 3 days for extraction and then 3 days for analysis.

(3) To ensure that iodide was not converted to iodate, every fifteen samples were analyzed for iodate.

(4) C: Critical, NC: Non-critical

(5) Al, Sb, Ba, Be, Cd, Ca, Cr, Co, Cu, Fe, Mg, Mn, Mo, Ni, K, P, Ag, Na, and Zn were analyzed by Method 6010B and Hg by Method 7470A.

(6) ALSI filtered/preserved the DOC samples from 2 x 40 mL unpreserved VOA vials.

SECTION 5
PRE-DEMONSTRATION INVESTIGATION

This section describes the use of a conservative tracer, bromide, to investigate the hydraulic properties of the demonstration plots as well as the reliability of the SPRD tracer injection system to deliver a precise low volume, high concentration of the tracer during the long-term application. After various rounds of bench top and field rate measurements ranging from 4 to 12 hours in duration, on February 1, 2001, in accordance with a temporary permit granted from California Water Quality Control Board, SPRD conducted a bromide injection test for approximately one month. The setup and startup of the test was led by SPRD personnel and the subsequent sampling and daily inspection of the tracer circulation and injection system was conducted by the NBVC staff. In order to determine the breakthrough curve as early as possible, two vials were collected from each sampling location; one was analyzed on site using an ion-specific probe, while the second was shipped to the SPRD laboratory in Ada, Oklahoma, for analysis. Daily probe readings were used by the SPRD TPM to estimate breakthrough curve at various transects in the Test and Control plots.

5.1   Bromide Tracer Test

The demonstration stakeholders agreed to implement a tracer circulation and injection system equipped to operate in a passive mode, thereby mimicking the site's natural conditions as closely as possible. That is, the system was able to deliver a small volume of concentrated tracer solution under a natural gradient through multiple injection points which were spaced in proximity with two rows to induce a curtain in the aquifer. Since the system operated under natural conditions, there was no change in the hydraulic head distribution or geochemistry of the demonstration plots. Therefore, the information generated was a true representation of the transient ground water.

To establish the background water quality parameters at the initiation of the test, the demonstration showed that the plots were anaerobic with DO measurements within the injection wells being less than 1 mg/L. The background bromide concentration was about 1 mg/L. During the course of the tracer test, the upgradient wells were also measured routinely for the determination of the background bromide concentration.

The test was formulated to add up to 10 mg/L, the amount permitted by the State of California, of bromide ion to the receiving aquifer. Sodium bromide was used in the test which was initiated on February 1, 2001.

As discussed in Section 2-3, the pre-demonstration bromide tracer test was successful in determining the extent of continuity between the 38 upgradient tracer injection wells, oxygen and propane sites, and downgradient observation wells, as well as the efficiency of the injection wells. It was also successful in defining the relative permeability of vertical and horizontal flow paths through the Test and Control Plots. The criteria selected for demonstrating the appearance of bromide at a monitoring point above background was 2.0 mg/L.

The first observation based on the results of the pre-demonstration bromide test was that the injection wells operated as designed. This was supported by significant bromide occurrences in both the nearby oxygen and propane injection wells. In all of the 8 oxygen injection wells in the Test Plot and 6 of the 8 oxygen injection wells in the Control Plot, bromide concentrations in excess of 2.0 mg/L were detected. The propane injection wells in the Test Plot did as well with "hits" in all seven wells. It is significant in that these findings demonstrated that d-MTBE and tracers introduced by the injection wells would be in direct and constant communication with the vendor treatment gases. The tracer injection system delivered the proper concentration consistently and reliably for the duration of the injection period. Based on the results of the pre-demonstration study, there was sufficient evidence that the system would operate adequately to complete the QAPP.

The second observation concerns continuity between the injection wells and downgradient observation wells in the Test and Control Plots. Although most bottom-screen wells received bromide concentrations above the 2.0 mg/L concentration, the variation was significant. It was also determined that bromide tracer activity in the middle and upper zones was very limited. This finding was reflected by generally low concentrations of intrinsic MTBE, suggesting that the natural ground-water flow at these locations was limited.

It was important to characterize flow paths through the Test and Control Plots, particularly with respect to the design of the sampling plan. One approach was to characterize the relative time of travel by determining the initial tracer breakthrough time between the injection wells and downgradient points of observation as shown in Table 5-1. The distances downgradient from the injection wells to observation points are provided in Table 5-1 only to obtain a relative sense of hydraulic conductivity. Since the times provided are initial breakthrough values, they cannot be used to calculate the true ground-water velocities.

Table 5-1
Initial Breakthrough Periods for
Downgradient Observation Points

Control Plot			Test Plot		
Location	Distance Feet	Breakthrough Days	Location	Distance Feet	Breakthrough Days
OIPC1	3	ND	OIPT1	3	20
OIPC2	3	8	OIPT2	3	19
OIPC3	3	6	OIPT3	3	26
OIPC4	3	13	OIPT4	3	13
OIPC5	3	6	OIPT5	3	6
OIPC6	3	6	OIPT6	3	13
OIPC7	3	14	OIPT7	3	4
OIPC8	3	ND	OIPT8	3	17
			PIPT1	6	48
			PIPT2	6	26
			PIPT3	6	19
			PIPT4	6	15
			PIPT5	6	29
			PIPT6	6	17
			PIPT7	6	19
C21B	8.5	7	T21B	8.5	35
C22B	8.5	9	T22B	8.5	9
C23B	8.5	13	T23B	8.5	33
C24B	8.5	26			
C31B	11	26	T31B	11	41
C32B	11	13	T32B	11	29
C33B	11	26	T33B	11	41

Note: ND – Non-detect.

The data in Table 5-1 demonstrate the heterogeneity of the various flow paths and that tracer breakthrough times in the Control Plot are lower than those in the Test Plot. However, the data further illustrate that the hydraulic communication between the injection wells, vendor's treatment gases, and downgradient treatment zone and observation points is adequate for evaluating the technology.

Another finding of importance to the design of the sampling plan was the vertical distribution and frequency of bromide detection over 2.0 mg/L. Between February 1, 2001, and March 30, 2001, 586 samples were analyzed for bromide in the Control Plot. Of these, 124 were 2.0 mg/L or greater in bottom screens and 19 were 2.0 mg/L or greater in the middle and upper screens with the most frequent of these being 9 at location C21Y. During the same period, 672 bromide analyses were made in the Test Plot of which 111 were 2.0 mg/L or greater in the bottom screens and only 4 were 2.0 mg/L or greater in the middle and upper screens.

The most significant finding of the pre-demonstration bromide tracer test was that ground-water flow was primarily confined to the bottom of the aquifer. This allowed the sampling plan for the evaluation demonstration to be expanded from the original 7 sampling events to 15 sampling events, within budget limitations, thereby strengthening statistical confidence in the projects results.

Although the bromide tracer injection was halted on February 28, 2001, the high frequency of sample collection continued until March 30, 2001. The purpose of this portion of the investigation was to observe the tracer to return to background. The sampling for bromide was conducted on two other episodes prior to the start of the long-term monitoring, once on May 30, and the second time on August 1, 2001. It is noted that for the duration of the test (February 28, 2001 – March 30, 2001), over 5000 samples were analyzed by ManTech, a SPRD on-site analytical contractor.

SECTION 6
TREATMENT EFFECTIVENESS - RESULTS

This section describes the results of Envirogen's demonstration evaluation at Port Hueneme National Environmental Test Site. Section 7 is organized to discuss and conclude the technology vendor's effectiveness against the project's objectives.

6.1 DEMONSTRATION OBJECTIVES AND APPROACH

The primary objective of the demonstration was to determine if biodegradation is occurring in the Test Plot to the extent that MTBE is remediated at or below 5 µg/L. The approach was to monitor Test and Control Plots for a period of 38 weeks (June 2001 – March 2002) and make appropriate analyses as outlined in the project PQA to determine if the objective had been met.

During the pre-demonstration investigation (February 1-28, 2001), all 38 injection wells were sampled 15 times for bromide analysis. Although there were differences between the bromide concentrations within the various injection wells, due in large measure to variations in hydraulic conductivity, the average concentrations within the Test and Control Plots were remarkably equal. The Test Plot had an average bromide concentration of 51.9 mg/L (STDEV = 10.8) while the Control Plot had an average concentration of 51.0 mg/L (STDEV = 10.9).

During the demonstration phase of the project the injection wells were also sampled frequently for iodide and d-MTBE. Although variations in concentrations between individual injection wells were observed, as during the bromide test, averages over the duration of the project were very similar. The average iodide concentration in the Test Plot injection wells was 30.2 mg/L (STDEV = 7.9) while that in the Control Plot was 31.2 mg/L (STDEV = 9.4). The results of d-MTBE sampling demonstrated similar behavior with an average concentration of 3,217 µg/L (STDEV = 382) in the Test Plot, while that of the Control Plot was 2,969 µg/L (STDEV = 188).

6.2 DEMONSTRATION PROCEDURES

Conservative tracers were introduced into the aquifer by a series of injection wells to determine aquifer flow paths in the Test and Control Plots. Deuterated MTBE (d-MTBE) was also introduced through the injection wells to avoid possible complications resulting from variations in the intrinsic MTBE. The evaluation procedure involved sampling the Control and Test Plots at upgradient and downgradient monitoring wells with screens located at the top, middle, and bottom of the aquifer. Analyses were also made to determine the presence of degradation (daughter) products, and geochemical parameters to evaluate changes expected due to the biodegradation processes.

6.2.1 MTBE Reduction

The major parameter in evaluating the technology demonstration is MTBE. Understanding its behavior before and during the tenure of the test is critical to evaluating the effectiveness of the enhanced in-situ bioremediation of MTBE. For example, samples taken from 5 monitoring wells near the east fence and 5 monitoring wells on the west fence surrounding the EPA plots on October 4, 2000, indicated that MTBE concentrations across the Test and Control Plots were varied, as shown in Figure 6-1.

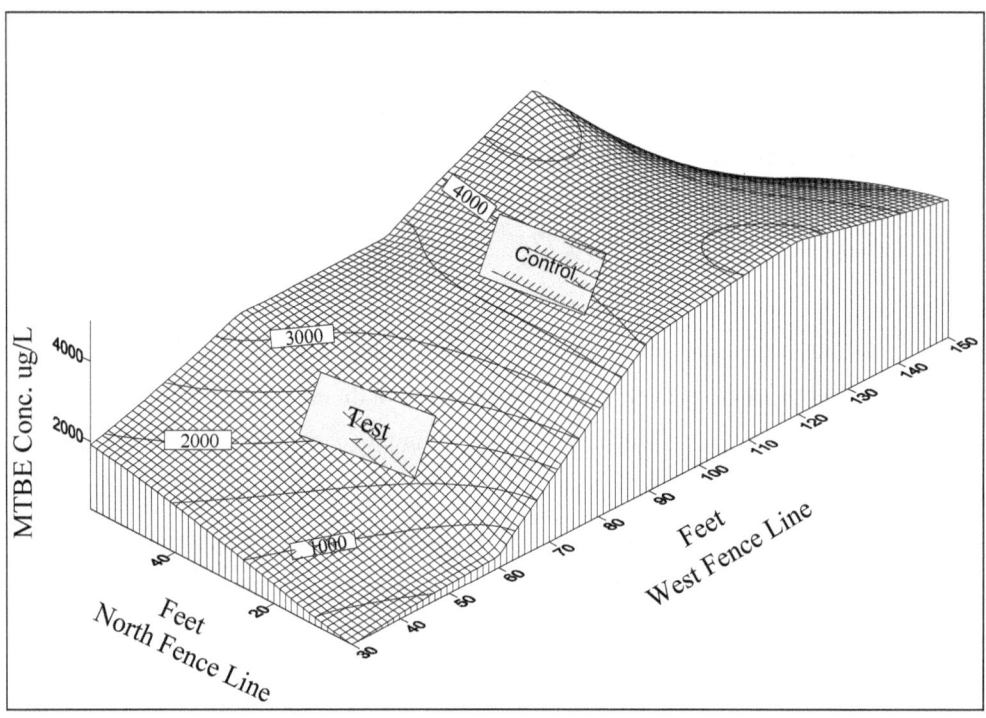

Figure 6-1. MTBE Concentration in the Vicinity of the Envirogen Site on October 4, 2000

On November 11, 2000, roughly five weeks later, the same wells were sampled again for intrinsic MTBE with the results shown in Figure 6-2. The MTBE concentration throughout the test area was reduced by about 500 µg/L. It is also noted that again, the Control Plot is roughly 2,000 µg/L MTBE higher than the Test Plot and that the downgradient area of the Test Plot is significantly lower in MTBE concentration than upgradient areas.

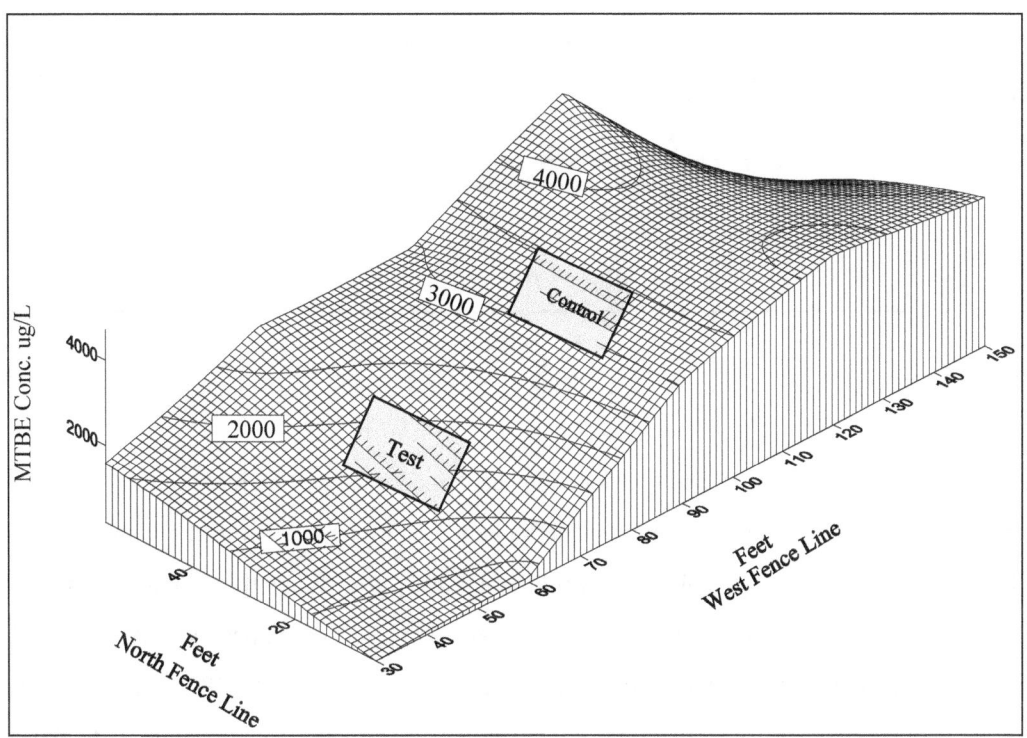

Figure 6-2. MTBE Concentration in the Vicinity of the Envirogen Site on November 11, 2000

This observation can be supported by comparing the average bottom screen MTBE concentrations in all of the upgradient and downgradient wells in the Test and Control Plots, as shown in Figure 6-3. This figure is comprised of both Envirogen and EPA data. For well locations, see Figure 3-1.

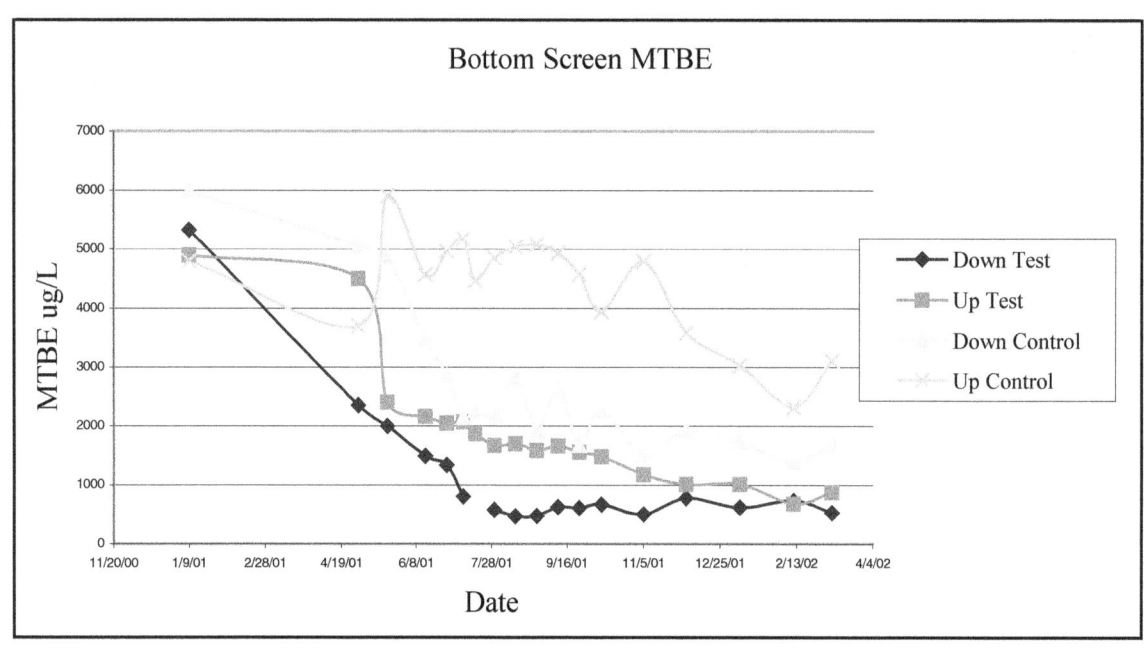

Figure 6-3. The Average Deep Screen MTBE Concentrations in the Test and Control Plots

As shown in Figure 6-3, on 01/09/01 the four locations contained relatively high concentrations including: downgradient Test Plot at 5,329 µg/L; upgradient Test Plot at 4,900 µg/L; downgradient Control Plot at 5,989 µg/L; and upgradient Control Plot at 4,800 µg/L. When the evaluation test period began on June 14, 2001, the downgradient Test Plot was at 1,493 µg/L; upgradient Test Plot at 2,160 µg/L; downgradient Control Plot at 3,471 µg/L; and upgradient Control Plot at 4,580 µg/L. The order of increasing intrinsic MTBE concentrations reflected in Figure 6-3 is downgradient and upgradient Test Plot followed by downgradient and upgradient Control Plot. This order is fairly well mirrored in Figures 6-1 and 6-2.

The injection wells in both the Test and Control Plots, which are upgradient from the treatment zone, were also sampled for MTBE throughout the period of the demonstration evaluation. The results from those analyses also confirm that nascent ground water entering the treatment zone continued to decrease during the evaluation. For example, in the Test Plot injection wells the average MTBE concentration decreased from 1,102 µg/L on July 18, 2001, to 280 µg/L on March 12, 2001. During this same period, the average MTBE concentration in the Control Plot injection wells decreased from 2,841 to 1,410 µg/L.

Variations in the intrinsic MTBE concentrations at the bottom monitoring well screens can also be depicted by downgradient columns such as T23B, T33B, T43B, T53B, and T62B as well as C23B, C33B, and C42B. For example, the average MTBE concentration for each sampling event at the three Test Plot downgradient columns, whose first wells are T21B, T22B, and T23B, are shown in Figure 6-4.

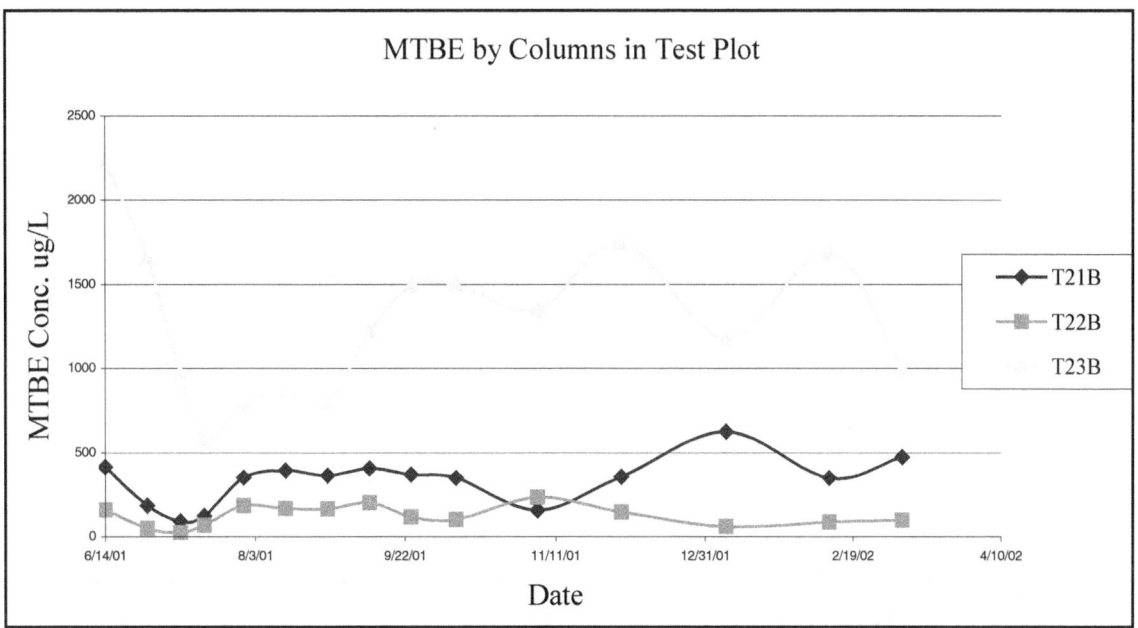

Figure 6-4. MTBE by Flow Paths (Columns) in Test Plot

Similarly, the average intrinsic MTBE concentration for each sampling event at the four Control Plot downgradient columns, whose first wells are C21B, C22B, C23B, and C24B, is shown in Figure 6-5.

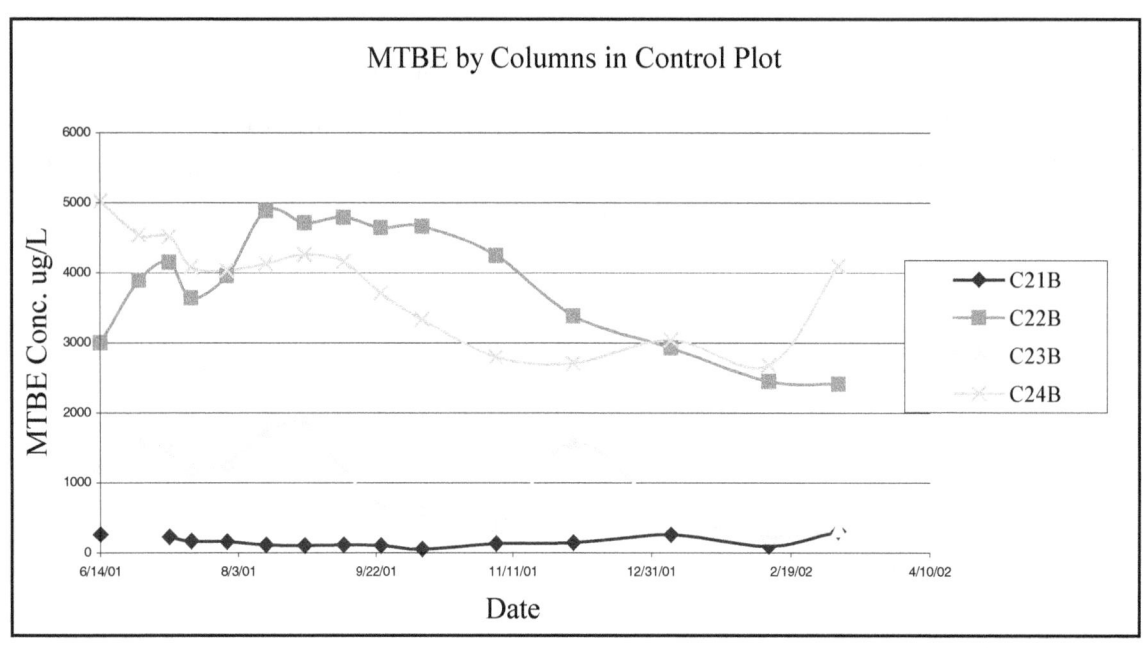

Figure 6-5. MTBE by Flow Paths (Columns) in Control Plot

6.2.2 d-MTBE Reduction

Although MTBE concentrations upgradient and downgradient of the treatment zone were evaluated in both the Test Plot and Control Plot, a focus of the evaluation technology was on the fate of deuterated-MTBE (d-MTBE) added to the system. d-MTBE was used as a non-conservative tracer in part to avoid possible uncertainties resulting from fluctuations in intrinsic MTBE concentrations. The use of d-MTBE also could provide evidence of biodegradation by tracking the generation of d-MTBE daughter products. For this demonstration the deuterated daughter products that were tracked included acetone-d6, 2-propanol-d6,d8, and *tert*-butyl alcohol-d9,d10. Although the presence of formaldehyde was tested, it was not possible to distinguish between deuterated and non-deuterated formaldehyde. Therefore, the generation of this daughter product will be representative of total formaldehyde.

Alterations in d-MTBE resulting from the demonstration can be considered in various ways by evaluating changes in concentration upgradient and downgradient in the Test and Control Plots, and in downgradient columns (i.e., T23B, T33B, T43B, T53B, T62B and C23B, C33B, and C42B) over the period of the 15 sampling events. Again, the evaluation of results is confined to the bottom screens because of the inactivity of the middle and upper screens.

54

For example, the average d-MTBE concentration for each sampling event at each of the three distinct flow paths in the Test Plot referred to as downgradient columns, whose first wells are T21B, T22B, and T23B, is shown in Figure 6-6. As demonstrated in other tracer tests, the T23B column is the most active followed by T22B and T21B. It should be pointed out that some of the fluctuation in d-MTBE averages is believed to be caused by the eight brief occasions when the injection system was not operating.

The following is a sketch of eight occurrences when the tracer injection system was inoperative:

- June 29, 2001 - down time was for 5 days only in the Control Plot due to operator failing to turn on the Control Plot metering pump.

- July 16, 2001 - down time was 5 hours due to a short pre-scheduled power outage for NBVC Port Hueneme Site and the performance of a 5-hour tracer injection flow-rate test.

- July 21, 2001 – down time 8 days Test Plot only due to operator failing to open the tracer reservoir valve.

- July 29, 2001 - down time was 3 hours due to a pre-scheduled power outage for NBVC Port Hueneme Site.

- November 26, 2001 - down time 24 hours caused by inability to locate two d-MTBE ampules which were available inside the refrigerator at the EPA shed.

- February 2, 2002 – down time 48 hours due to power outage caused by a storm.

- February 9, 2002 – down time 48 hours due to power outage caused by a storm.

- March 6, 2002 – down time 5 days because of power interruption at the Base during the weekend. However, the operator failed to turn on the metering pumps.

Although these unfortunate down times resulted in a short interruption of the tracer events, it was observed that the tracer wells cleared quickly indicating they were free from obstructions, and recovered quickly indicating that the tracer injection systems operated as designed.

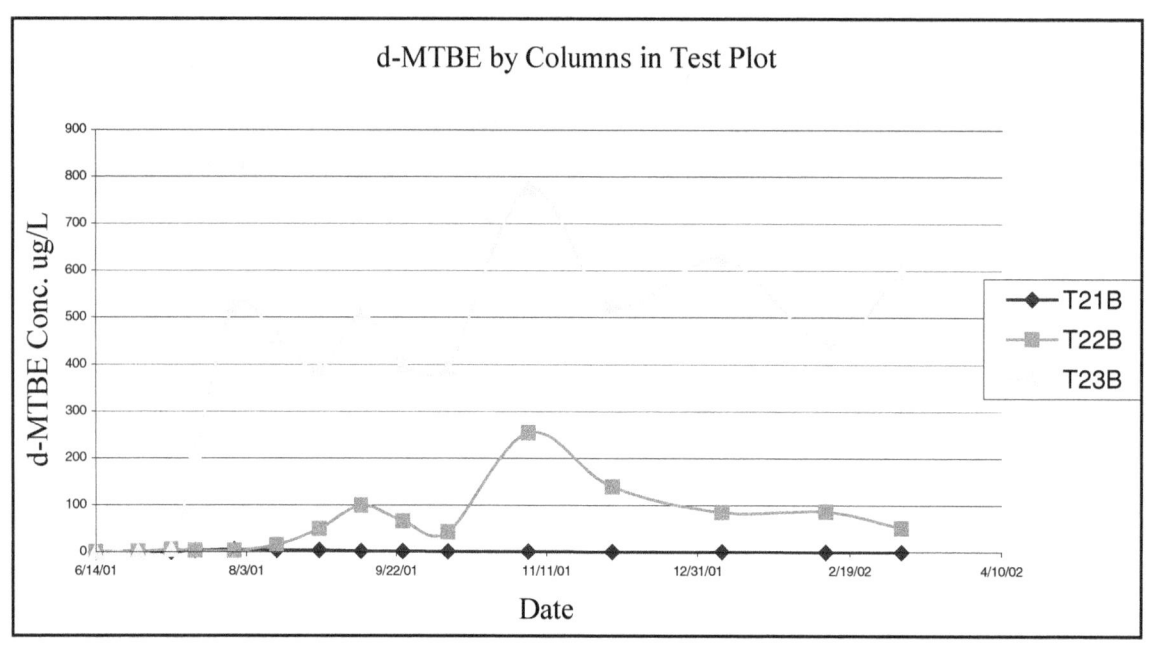

Figure 6-6. d-MTBE in Downgradient Columns of Test Plot

Similarly, the average d-MTBE concentration for each sampling event at the four Control Plot downgradient columns, whose first wells are C21B, C22B, C23B, and C24B, is shown in Figure 6-7.

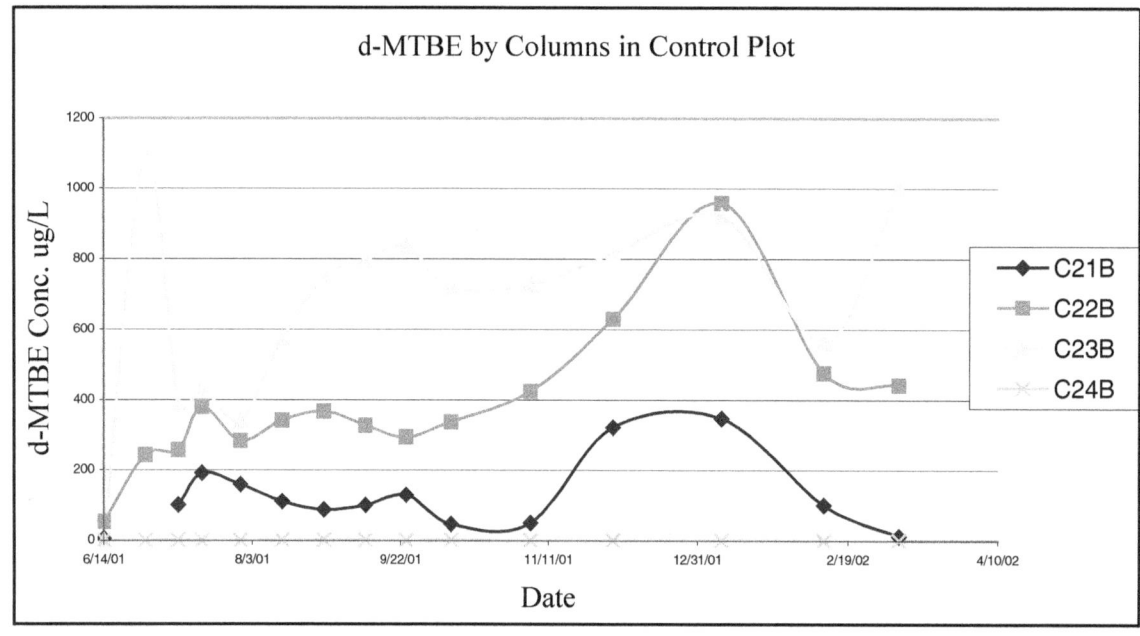

Figure 6-7. d-MTBE in Downgradient Columns of Control Plot

Again, as demonstrated by other tracer studies, the C23B column is the most active, followed by C22B, C21B, and C24B (Figure 6-7). In addition to evaluating d-MTBE concentration changes over time along flow paths in the Test and Control Plots, it is useful to compare the average of all downgradient bottom-screen d-MTBE concentrations over the 15 sampling events. In a way, this is like evaluating changes in the total downgradient d-MTBE mass over time in both the Control and Test Plots. As shown in Figure 6.8, the least squares fit in both the Control and Test Plots indicates that downgradient d-MTBE concentrations increased over the study period at about the same rate with those in the Control Plot being somewhat higher.

It was noted that d-MTBE was detected at low levels in the upgradient monitoring wells. Table 6-1 describes the extent to which d-MTBE was detected in the upgradient Control and Test Plots. It is noted that the determination of MTBE, d-MTBE was accomplished by the analysis of collected samples using GC/MS methodology with reporting limits (minimum quantitation limit) of 1 μg/L. Consequently, the numbers below 1 μg/L are estimated values and have no bearing on the evaluation of project objectives.

Table 6-1

Detection of d-MTBE in Upgradient Monitoring Wells

Control Plot			Test Plot		
Well	**Sample Event**	**Conc. μg/L**	**Well**	**Sample Event**	**Conc. μg/L**
C13B	6	0.7	T11B	2	0.19
	13	0.5		4	1.2
				6	0.2
C14B	3	0.4		7	0.3
	7	0.1		9	1.9
	13	0.4			
			T12B	9	0.5
			T13B	2	0.22
				4	0.7
				9	0.4
			T14B	4	0.4
				9	0.3
			T15B	9	0.2
			T16B	3	0.3
				4	0.2
				7	0.1
				9	0.2

Note:

Location and designation of well screens are provided in Figure 3-1.

Abbreviations:

C: Control
T: Test
B: Deep Screen

58

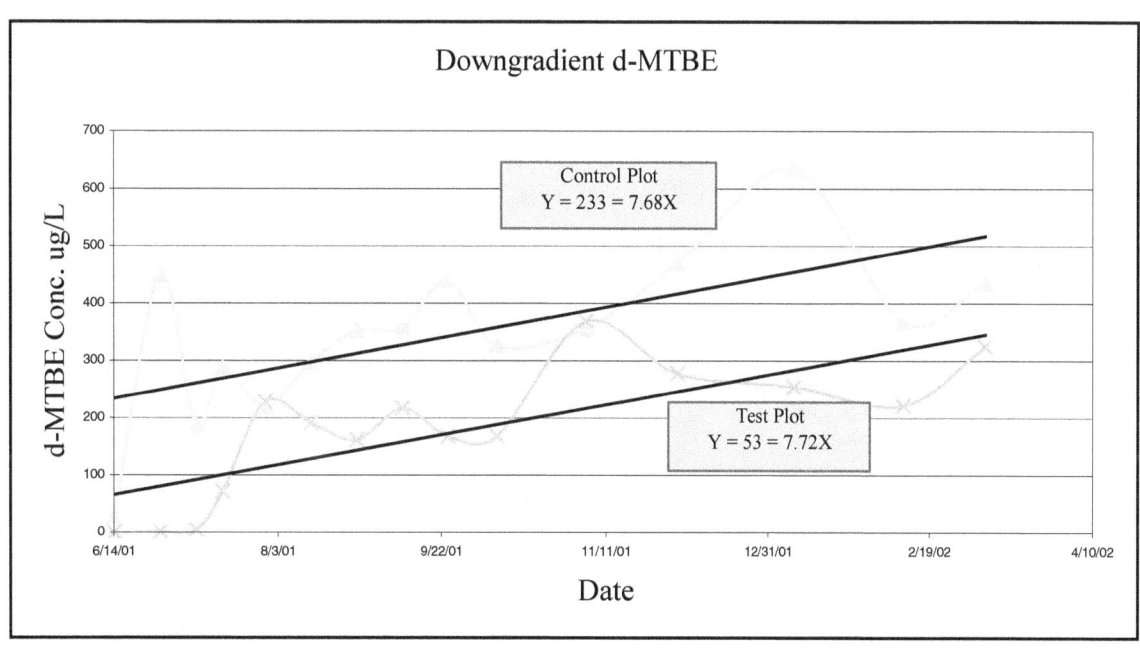

Figure 6-8. Downgradient d-MTBE Concentration in Control and Test Plots

6.2.3 Daughter Products

An indirect way to assess biodegradation processes is to determine the quantity of MTBE and d-MTBE daughter parameters produced which must be directly associated with the reduction in mass of the contaminant of concern. During the processes involved in the biodegradation of contaminants, degradation or daughter products are created. If the process proceeds to the mineralization endpoint, the daughter products themselves will be further remediated until only CO_2 and water remain.

During the demonstration evaluation, one would expect daughter product masses to be commensurate with MTBE and d-MTBE mass reduction to the extent that the remediation process was effective in the downgradient Test Plot. It would also be expected that the production of daughter products in the downgradient Test Plot would be significantly higher than in the downgradient Control Plot.

The results of the 15 sampling events are summarized in Table 6-2 for the Control Plot and Table 6-3 for the Test Plot. It should be noted that these data refer only to those samples collected at the deep screens since, as mentioned before, little activity was observed in the upper and middle zones of the aquifer.

Examples of detections in the middle and upper screens (not reflected in Tables 6-2 and 6-3) demonstrate that the highest average value of TBA over the 15 sampling events in the Control Plot was 111 µg/L (C12Y-middle screen) followed by 73 µg/L (C11Y-middle screen). The remainders of the screens were below 40 µg/L and most were non-detectable (ND). TBA concentrations in the Test Plot were generally lower with the highest average 15 sampling event value being 59 µg/L at T32Y (middle screen). In addition to TBA, low levels of other daughter products were detected at five medium and shallow screens. For example, at T21R 8.1 µg/L acetone (Event 15) and 6.8 µg/L 2-propanol (Event 12) were detected. Low levels of acetone were also detected at T31R (upper screen), T33Y (middle screen), T33R (upper screen), and T51R (upper screen).

As shown in Tables 6-2 and 6-3, which represents only bottom-screen values, many of the reported concentrations were very low (e.g., less than the practical quantitation limit and /or method detection limit). Therefore, in order to calculate the descriptive statistics presented in Tables 6-2 and 6-3, these low values were replaced with the method detection limit (MDL) for the parameters under consideration. These detection limits included acetone/d-acetone, 10 µg/L; 2-propanol/d-2-propanol, 4-20 µg/L; formaldehyde, 10 µg/L; *tert*-butyl alcohol-d9 and -d10, 10 µg/L; *tert*-butyl alcohol, 4 µg/L.

Three Injection Wells in the Test Plot (P1, P8, P16) and in the Control Plot (P22, P28, P32) were also sampled and analyzed for suspected daughter products. Although these locations are upgradient from the treatment zone in each plot, the d-TBA average concentration over the 15 sampling events for these injection wells was 125 µg/L in the Test Plot and 168 µg/L in the Control Plot.

Table 6-2
Daughter Products in Control Plot

Target Analytes	Upgradient Control Plot				Downgradient Control Plot			
	Mean	STDV	High	Low	Mean	STDV	High	Low
Acetone	N.D.				N.D.			
Acetone-d6	N.D.				N.D.			
2-Propanol	N.D.				N.D.			
2-Propanol-d6,d8	N.D.				N.D.			
Formaldehyde	N.D.				N.D.			
t-Butyl Alcohol-d9,d10	N.D.				25	10	237	N.D.
tert-Butyl Alcohol	105	29	399	13	63	46	348	N.D.
All values are µg/L								
Average of bottom screens over 15 sampling events.								

Table 6-3
Daughter Products in Test Plot

Target Analytes	Upgradient Test Plot				Downgradient Test Plot			
	Mean	STDV	High	Low	Mean	STDV	High	Low
Acetone	N.D.				10		11	N.D.
Acetone-d6	N.D.				N.D.			
2-Propanol	N.D.				N.D.			
2-Propanol-d6,d8	N.D.				N.D.			
Formaldehyde	N.D.				N.D.			
t-Butyl Alcohol-d9,d10	N.D.				29	13	215	N.D.
tert-Butyl Alcohol	29	9	97	N.D.	29	29	290	N.D.
All values are µg/L								
Average of bottom screens over 15 sampling events.								

Abbreviations:
STDV: Standard Deviation   N.D.: Non Detect   µg/L: microgram per liter

Note: The target analytes detection limit and PQLs are described within the text. See Section 6.2.3.

6.2.4 Water Quality Measurements

An indirect approach in evaluating remediation effectiveness is the assessment of alterations to ground-water geochemistry resulting from biodegradation processes. Both the bacterial requirements for growth and respiration, as well as degradation products, alter many intrinsic geochemical parameters. The requirements include nutrients and sources of energy while the degradation products include mineralization end points.

In this case the assessment can be made between upgradient and downgradient parameter concentrations within the Test Plot as well as comparisons between the Test Plot and the Control Plot. The following parameters were selected to reflect geochemical alterations that might occur during the demonstration.

- Alkalinity is expected to increase due to the production of mineralization end products including carbonate ions;
- Electron donors and nutrients including ammonia-nitrogen, nitrate and nitrite, phosphorus, orthophosphate, and sulfate are expected to be reduced in concentration after utilization by microbes; and
- Total and dissolved organic carbon (electron donors). Usually the mass of the electron donor compounds needed to stimulate bacterial growth is necessarily much larger than the mass of the contaminant being degraded.

The results of the ground-water geochemical analysis are presented in Table 6-4 for the Control Plot and Table 6-5 for the Test Plot. The information is comprised of the average parameter concentrations at the bottom screens as determined by the results of the 15 sampling events. The information is further divided into upgradient and downgradient locations with respect to the treatment transects in each plot.

Table 6-4
Water Quality Measurements in Control Plot

Target Analytes	Upgradient Control Plot				Downgradient Control Plot			
	Mean	STDV	High	Low	Mean	STDV	High	Low
Alkalinity	533	22	558	519	502	31	544	452
Ammonia-Nitrogen	0.55	0.24	0.79	<0.3	0.56	0.21	0.76	<0.3
Nitrate+Nitrite	0.13	0.14	0.29	<0.05	0.10	0.12	0.44	<0.05
Phosphorus, Total	0.1	0.12	0.24	<0.02	0.2	0	0.2	0.2
TOC	3.3	0.1	3.4	3.2	3.5	0.4	4.1	2.8
TOC Dissolved	3.0	0.1	3.1	3	3.3	0.3	3.8	2.7
Nitrate	0.3	0.23	0.48	<0.05	0.22	0.28	0.69	<0.05
Nitrite	0.32	0.04	0.37	<0.05	0.26	0.22	0.59	<0.05
Orthophosphate	0.02	0	0.02	<0.02	0.02	0	0.02	<0.02
Sulfate	1132	101	1241	1041	1167	37	1217	1134
All values are mg/L								
Average of Bottom Screens Over 15 Sampling Events.								

Table 6-5
Water Quality Measurements in Test Plot

Target Analytes	Upgradient Test Plot				Downgradient Test Plot			
	Mean	STDV	High	Low	Mean	STDV	High	Low
Alkalinity	485	22	509	451	437	34	478	475
Ammonia-Nitrogen	0.89	0.07	0.97	0.82	0.57	0.28	1.08	<0.3
Nitrate+Nitrite	0.06	0.05	0.14	<0.05	0.16	0.15	0.51	<0.05
Phosphorus, Total	0.08	0.06	0.13	<0.02	0.02	0.0	0.02	<0.02
TOC	3.3	0.2	3.6	3.1	3.6	0.3	4.4	2.8
TOC Dissolved	3.0	0.3	3.6	2.8	3.3	0.3	4.2	2.7
Nitrate	0.05	0.0	0.05	<0.05	0.3	0.3	1.17	<0.05
Nitrite	0.56	0.3	0.81	<0.05	0.1	0.1	0.59	<0.05
Orthophosphate	0.02	0.0	0.02	<0.02	0.02	0.0	0.02	<0.02
Sulfate	1179	118	1300	1016	1189	119	1217	1103
All values are mg/L								
Average of Bottom Screens Over 15 Sampling Events.								

Abbreviations:
STDV: Standard Deviation   mg/L: milligram per liter

Note:
The numbers above detection limit and below practical quantitation limit are reported as less than (<).

As shown in Tables 6-4 and 6-5, many of the reported concentrations were very low. In addition, parameter concentrations were often reported as less than practical quantitation limit (<PQL) or non-detect (ND) values. In order to calculate the descriptive statistics, these values were replaced with the detection limit for the parameters under consideration.

6.3   WATER LEVEL MEASUREMENTS

Part of the sampling activities at the Port Hueneme MTBE remediation evaluation project have included the determination of aquifer water levels within and adjacent to both the Test and Control Plots at the site. During each sampling event, water level elevations were determined concurrently with geochemical parameters. During these events, observations were made concerning water spouts at the surface through monitoring wells. These events were described as well as photographed. As a result of these observations, during the 12th sampling event, real-time water level measurements were made to determine the effect of periodic oxygen injection pulses on flow characteristics in the vicinity of the oxygen injection transects in both the Test and Control Plots.

Water level readings were made on three dates: December 5, 2001, December 6, 2001, and December 7, 2001. Water levels were recorded using hand measurements and various transducers (i.e., Level Logger, Minitroll, and Transducer). Plots of static water levels on all dates show low relief, i.e., difference between highest and lowest elevations, ranging from 0.65 to 0.11 feet (7.8 to 1.3 inches) (19.8 cm to 3.3 cm). The December 6 and 7 plots show correspondence to regional ground-water table flow direction in the area.

December 5 plots during the first sparging event (time = 11:53:47 and 11:58:17) show ground-water mounding in the area, with one mound centered on MW PIPT-4, and the highest mound in the upgradient of MW OIPT-1 and MW OIPT-2.

Figures 6-9 through 6-12 provide examples of the piezometric surfaces for the Test and Control Plots under static conditions, as well as the changes induced during injection of the gases. As shown, water levels were elevated over 4 feet (1.21 m) in the Test Plot and about 3 feet (0.91 m) in the Control Plot. As shown in Figure 6-13, the injection of gases also causes water spouts at the surface through monitoring wells.

Although these sparging events only last about 5 minutes and occur only 4 times a day, the natural ground-water gradient of about 0.002 is increased orders of magnitude resulting in the detection of tracers in the upgradient monitoring wells and causing considerable disruption to the natural ground-water flow field and dispersal of the injected tracers. For example, the water in the injection wells was anaerobic before the project began, and contained high concentrations of oxygen thereafter. Shortly after the termination of the injection of gases, the oxygen concentration dropped significantly. For example, DO measurement of the injection wells during the last QAPP sampling event (March 11, 2002) indicated that 14 out of the 19 Test Plot injection wells had concentrations >15 mg/L and 7 out of 19 injection wells within the Control plots had concentrations of >15 mg/L. The last oxygen measurements conducted on April 30, 2002, indicated that 10 out of the 19 Test Plot injection wells were below 3 mg/L and 11 out of the 19 Control Plot injection wells were below 1 mg/L. In any event, these mounds caused considerable disruption to the natural ground-water flow paths.

December 5, 2001 Water Levels, Static Levels Test Plot

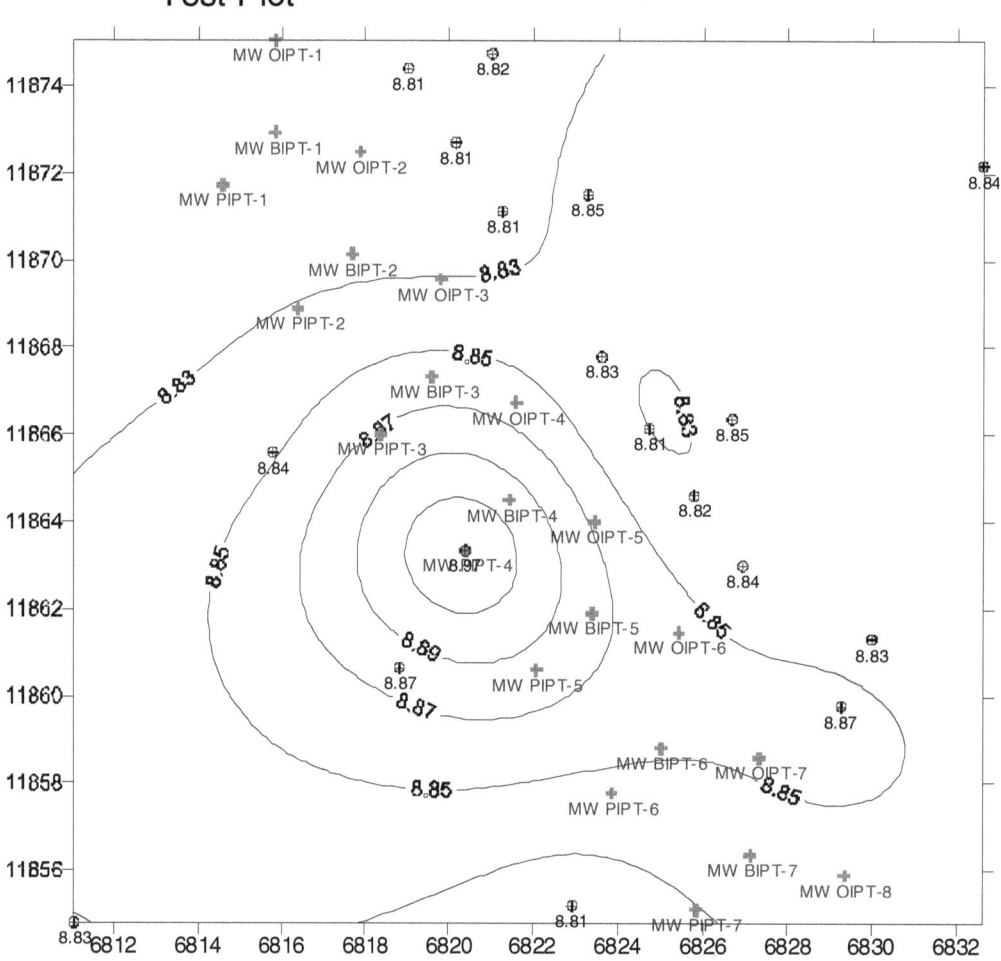

Figure 6-9. Static Water Levels in Test Plot

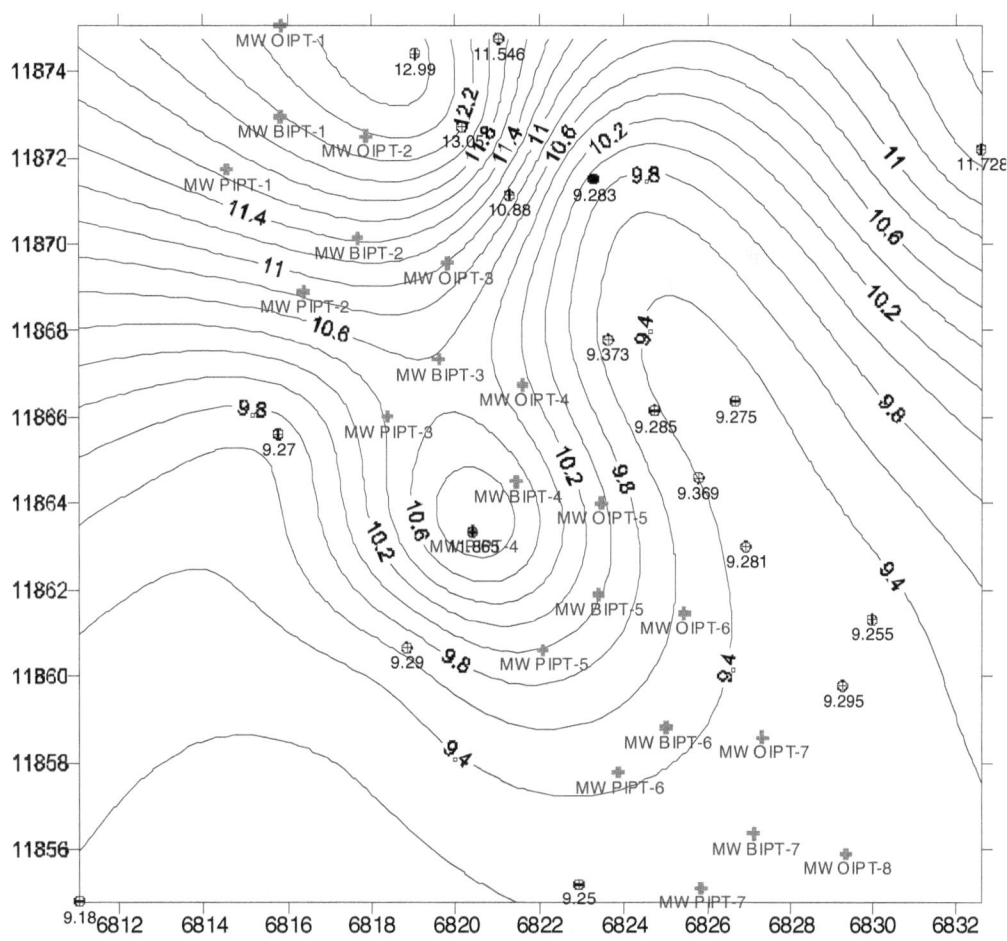

Figure 6-10. Maximum Water Levels in Test Plot

December 6, 2001 Water Levels, Static Levels Control Plot

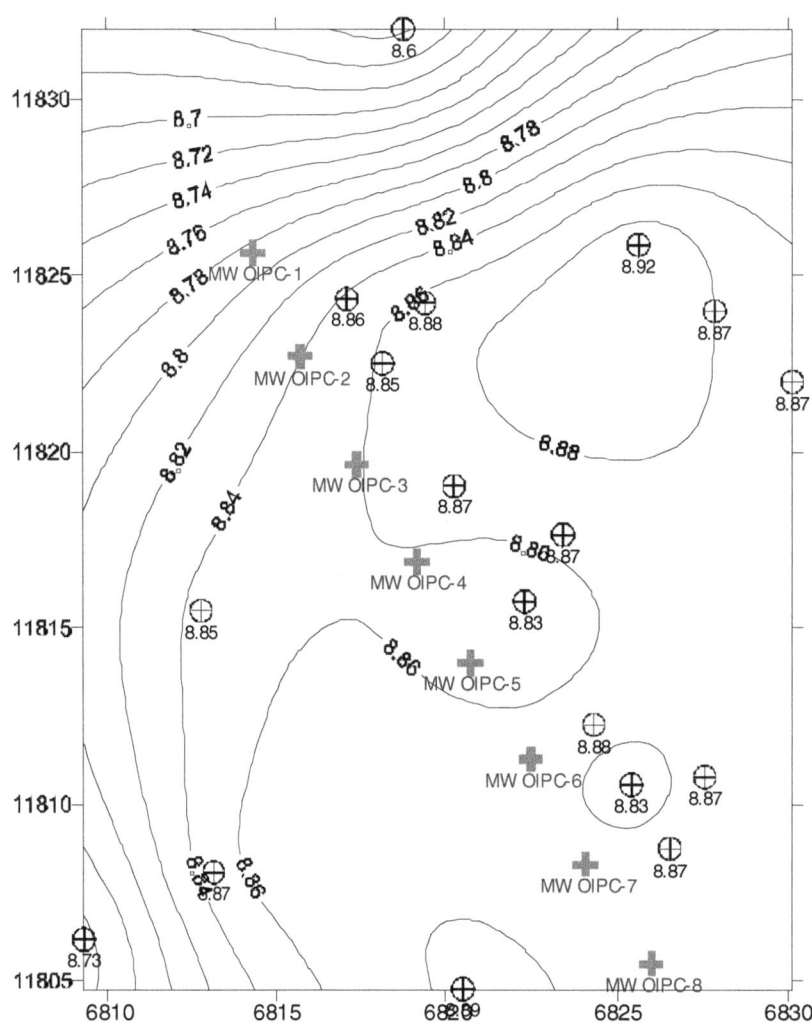

Figure 6-11. Static Water Levels in Control Plot

December 6, 2001 Water Table Time = 09:17:30 Maximum Water Levels During 1st Sparge Event Control Plot

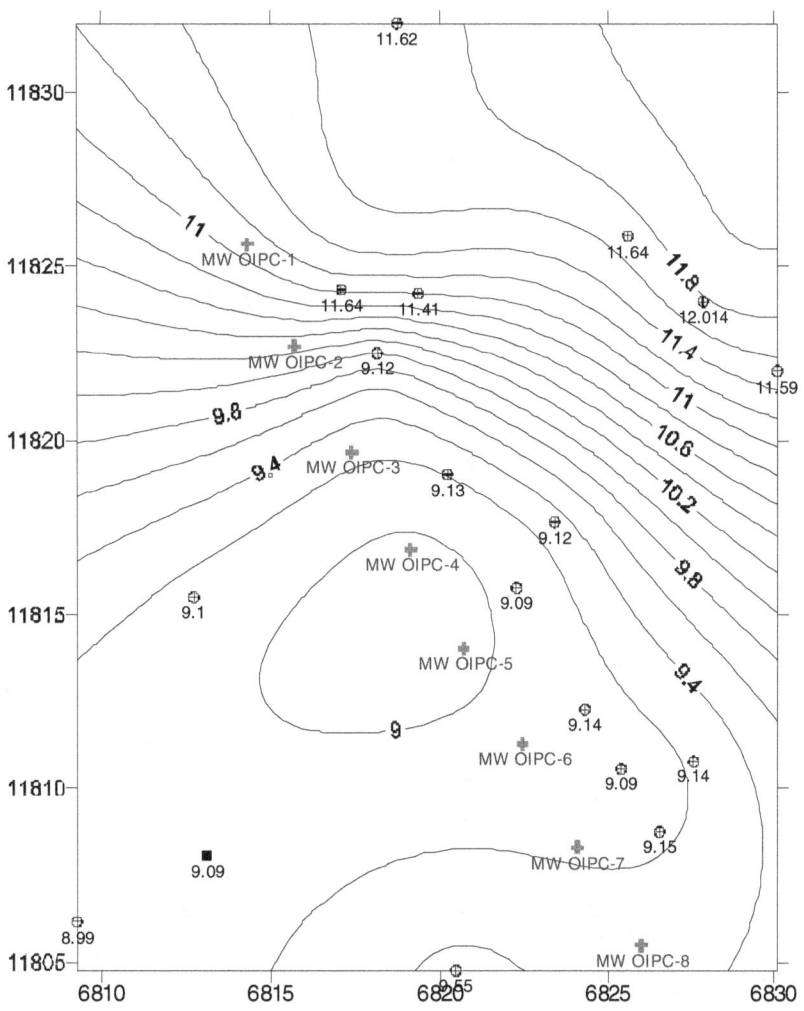

Figure 6-12. Maximum Water Levels in Control Plot

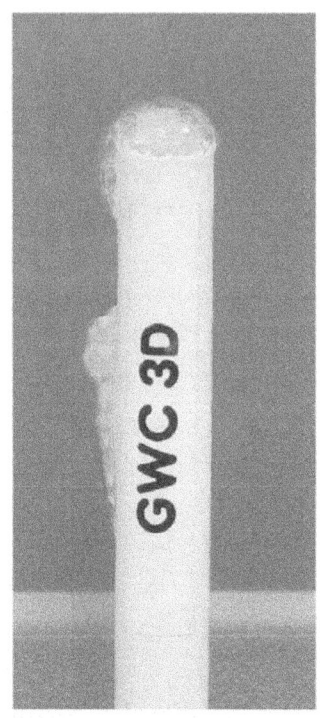

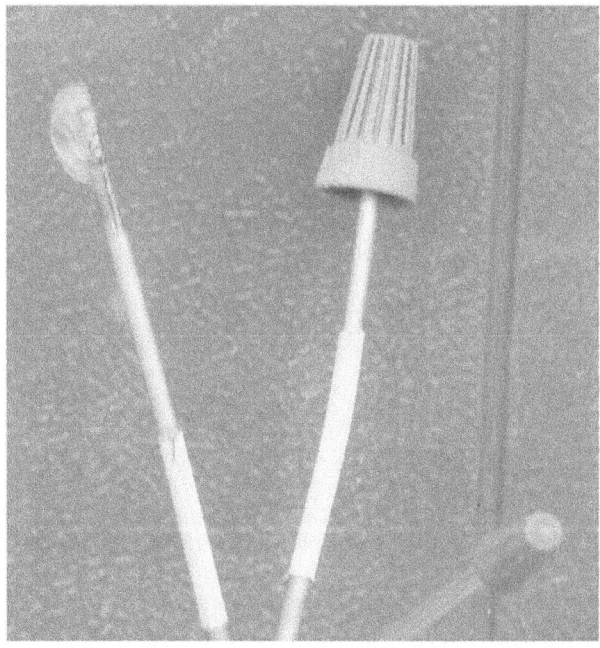

Figure 6-13. Pictures of Water Spouts at the Surface Through Monitoring Wells. GWC 3D - Envirogen deep screen in downgradient center of first transect in Control Plot (top). T14M - EPA middle screen in upgradint transect in Test Plot (bottom).

SECTION 7

TREATMENT EFFECTIVESNESS - CONCLUSIONS

This section addresses the statistical analysis of the results of the demonstration of the Envirogen technology at the NBVC in Port Hueneme, California, and describes the effectiveness of the technology in treating ground water contaminated with MTBE and other gasoline constituents. The technology demonstration was implemented and evaluated in one phase at the Middle Zone within the MTBE plume over a ten-month period, as shown in Table 2-4.

The EPA performed an independent evaluation of Envirogen's propane biostimulation technology through a joint effort between the National Risk Management Research Laboratory's (NRMRL) Subsurface Protection and Remediation Division (SPRD) and the Innovative Technology Evaluation Program. Although SPRD led the technical design and the performance of the evaluation process, each of the project stakeholders approved the SPRD-developed Work Plan entitled "Performance Monitoring of Enhanced In-Situ Bioremediation of MTBE in Ground Water, 2000," and reviewed its Quality Assurance Project Plan, SAIC, 2001, companion document. California Water Quality Control Board (WQCB) granted a project permit through a public hearing and NFESC and SPRD worked cooperatively to staff the field sampling crews and to manage the evaluation. Prior to the implementation of the field demonstration, the project PQA was approved by Envirogen (May, 2001).

7.1   BACKGROUND

As discussed in the previous sections (Section 5 and 6), the conceptual model to approach the evaluation process was to employ ground-water tracers and a surrogate tracer to determine the efficacy of the technology performance. In response, the pre-demonstration bromide tracer study was implemented under a passive system to inject low volume, high concentration of a bromide tracer solution under the natural gradient and anaerobic conditions prior to the operation of the Envirogen technology. The essence of the tracer study was to evaluate the intrinsic flow velocity, and communication of the various components of the system including injection wells, downgradient monitoring wells and the vendor's treatment gases injection points.

The second segment of the tracer injection, carried out during the evaluation process, was conducted under an active aerobic system resulting from the injection of treatment gases. The essence of the study was to avoid the influence of mounding and diversion potentials and only select the monitoring points

71

that are intercepted by the flow lines. The selection of d-MTBE as a surrogate for intrinsic MTBE was made to reduce the background fluctuations since biodegradation processes have the potential to be sporadic in nature. Similarly, in order to avoid the potential chaotic behavior that may result from air sparging, the selection of a conservative tracer alleviates the screening process thereby marking the flow paths between the tracer injection system and monitoring wells downgradient from the treatment gases. It should be noted, however, that since the tracer studies were representative of two distinct environmental conditions, passive and anaerobic as well as active and aerobic, their results cannot be readily compared.

7.2   PERFORMANCE ANALYSIS

The project participants agreed that the main focus of the EPA evaluation should be to determine the behavior of intrinsic MTBE and the tracer d-MTBE. The latter involved the iodide/d-MTBE ratio discussed in Section 2.3.1, which was proposed to differentiate between abiotic and biotic reduction, as well as provide the definition of qualified wells for analyses.

Of considerable importance in addressing these issues is the performance of selected conservative and non-conservative tracers. The Work Plan identified the use of iodide as a conservative tracer in the demonstration evaluation phase of the project because of its low background of approximately 20 μg/L as compared with bromide of approximately 1 mg/L, to avoid bromide residuals that could be present following the pre-demonstration tracer test, and its widespread application as a ground-water conservative tracer.

During the evaluation phase of the project, as was evidenced and previously described in the bromide pre-demonstration results (Section 5), intrinsic MTBE and d-MTBE fluctuated in time and locations within the Control and Test Plots. Also, the behavior of the iodide tracer, with respect to characteristics of the earlier bromide tracer (which increased in concentration throughout the test period) was evidenced in essentially four ways:

1.   Iodide appeared to duplicate the earlier bromide results,
2.   Iodide had a protracted delay in appearance in downgradient wells compared to that of bromide,
3.   Iodide would increase in concentration followed by a concentration reduction, and,
4.   Iodide remained at undetected levels while the earlier bromide concentrations attained and remained at various concentrations until being reduced after injection ended.

In order to determine the presence of iodate, the most abundant species in brackish water, as was stipulated by the QAPP, during the first three sampling events for iodide every sample was also analyzed for iodate but it was never detected. This frequency was then reduced to 1 out of every 15 samples during sampling events 4-15 with no detection.

It was also of interest to insure that the presence of other chemical species of iodide were determined. Therefore, selected samples from the locations which historically have shown high bromide concentrations were submitted to a laboratory to be analyzed for total iodine by ICPMS. In order to confirm the stability of the samples, the same samples were reanalyzed for iodide and iodate. The results have shown that (1) all of the samples were non-detect for iodate, (2) the same concentrations of iodide resulted from the reanalyzed samples, and (3) a good correlation was evident between iodide and total iodine.

In an attempt to determine the extent of ground-water flow alteration by the treatment system, a second bromide tracer test was started on October 29, 2001, which was carried out according to the pre-demonstration bromide tracer specifications. When compared, the October 29 tracer results appeared to have some of the same inconsistencies as those encountered when using iodide.

As was pointed out earlier, the original bromide tracer test was carried out prior to the initiation of the remediation demonstration when the natural aquifer system was not disturbed by treatment gases, as discussed in Section 4. It may be possible that this change in the natural flow system contributes to alterations in tracer behavior. In support of this argument, it appears that 8 to 10 of the bottom screen wells increased in iodide concentration after the treatment gases had been turned off. Furthermore, the injection of iodide continued until May 30, 2002, which was over 2 ½ months after the treatment system was discontinued.

7.2.1 Qualified Monitoring Points

The approved Pre-Quality Assurance Project Plan Agreement (PQA), May 2001, documented the statistical justification and confidence levels associated with the determination of the number of critical samples for the analytes of concern. Project participants agreed that these samples represented the experimental units for the evaluation of the primary objective.

To that end, each screen where iodide was detected at 500 μg/L or greater (qualified monitoring well) was located along with the corresponding d-MTBE concentration. This information for both the Control and Test Plots is provided in Table 7-1. It was determined that 93 qualified wells were in the Control Plot and 54 were in the Test Plot.

7.2.2   Statistical Analysis of Results

The objective of this data examination is to perform a statistical analysis on the demonstration evaluation d-MTBE data to determine if the bioremediation process (injecting propane, oxygen, and propane oxidizing bacteria) has reduced d-MTBE concentrations in the treatment plot. The effectiveness of the demonstration is determined by evaluating whether d-MTBE levels in the downgradient well samples of the Test Plot are at or below 5 μg/L over a 10-month period using one-sided hypothesis test on mean (80% UCL). This degradation is established by measuring d-MTBE concentrations in the "qualified samples" and determining whether, with 80% confidence, the estimate of the population mean is at or below 5 μg/L.

Data Analysis

Iodide provides evidence of ground-water flow in downgradient sampling wells. d-MTBE data are evaluated only for "qualified" monitoring points where iodide concentrations are greater than the practical quality detection limit, 500 μg/L. As specified in the PQA, sampling data from events 4 through 15 are used for evaluation. Based on this data selection criteria, there are 93 and 54 d-MTBE "qualified samples" in the Control Plot and in the Test Plot, respectively (Table 7-1). The d-MTBE data are neither described by normal distribution nor by log-normal distribution (Figures 7-3 and 7-4). However, the square-root transformed d-MTBE [sqrt(d-MTBE)] are well described by normal distribution (Figures 7-1 and 7-2). Therefore, the results in the following are obtained from the statistical analysis of square-root transformed data of d-MTBE.

TABLE 7-1
Qualified Monitoring Wells

	CONTROL					TEST			
Date	Days	Location	Iodide	d-MTBE	Date	Days	Location	Iodide	d-MTBE
28-Jun	21	C22B	1.92	243					
		C23B	2.70	1100					
9-Jul	32	C22B	2.54	258					
17-Jul	40	C22B	2.69	258	17-Jul	40	T23B	5.69	385
		C23B	2.36	598			T33B	1.04	30
		C32B	4.70	501					
		C33B	0.67	268					
30-Jul	53	C22B	3.25	206	30-Jul	53	T23B	9.50	1150
		C23B	3.10	488			T33B	2.72	410
		C32B	4.22	322			T43B	1.01	15
		C33B	0.65	283					
		C41B	4.29	325					
		C42B	1.24	248					
13-Aug	67	C22B	3.23	291	13-Aug	67	T23B	7.18	972
		C23B	4.97	703			T33B	1.47	559
		C31B	0.61	138			T43B	1.75	150
		C32B	3.94	374			T53B	2.55	152
		C33B	1.94	509					
		C41B	4.45	360					
		C42B	2.72	510					
27-Aug	81	C22B	2.15	236	27-Aug	81	T23B	0.75	134
		C23B	6.99	876			T33B	0.87	185
		C32B	3.93	396			T43B	2.10	488
		C33B	3.21	657			T52B	2.63	110
		C41B	5.10	468			T53B	2.55	620
		C42B	3.71	712			T62B	1.93	521
10-Sep	94	C21B	0.54	49	10-Sep	94	T33B	3.03	494
		C22B	3.13	253			T43B	3.06	515
		C23B	6.24	787			T53B	3.70	713
		C31B	0.61	150			T62B	1.84	733
		C32B	3.59	303					
		C33B	4.30	702					
		C41B	4.79	425					
		C42B	4.73	899					
		C21Y	0.68	101					
		C41Y	1.58	352					
		C22Y	5.08	1420					
24-Sep	109	C21B	0.62	72	24-Sep	109	T23B	2.20	115
		C22B	2.29	144			T33B	0.99	246
		C23B	7.98	840			T43B	4.04	487
		C31B	0.64	129			T53B	2.47	561
		C32B	3.06	251			T62B	1.26	595
		C33B	6.80	756					
		C41B	6.11	487					
		C42B	2.72	911					
8-Oct	123	C21B	0.51	32	8-Oct	123	T23B	2.47	198

Date	Day	Well	Iodide	d-MTBE		Date	Day	Well	Iodide	d-MTBE
		C22B	2.30	190				T33B	2.47	343
		C23B	5.63	612				T43B	2.35	322
		C31B	0.54	62				T53B	2.69	482
		C32B	2.68	238				T62B	1.47	622
		C33B	4.83	732						
		C41B	6.56	582						
		C42B	4.00	813						
		C22Y	9.56	1550						
		C41Y	1.29	213						
5-Nov	151	C21B	0.61	36		5-Nov	151	T23B	6.21	571
		C22B	2.25	135				T33B	3.71	678
		C23B	9.78	986				T43B	2.29	848
		C31B	0.71	64				T52B	0.75	552
		C32B	5.78	521				T53B	2.39	985
		C33B	1.05	695				T61B	1.29	339
		C41B	6.97	617				T62B	3.37	796
		C42B	1.61	496						
3-Dec	179	C21B	1.31	235		3-Dec	179	T23B	4.15	538
		C22B	8.26	581				T33B	2.51	898
		C23B	11.78	1210				T43B	1.46	162
		C31B	1.10	128						
		C32B	12.80	811						
		C33B	5.18	755						
		C41B	8.55	499						
		C42B	8.58	483						
		C22Y	1.63	532						
7-Jan	214	C22B	9.98	838		7-Jan	214	T23B	7.72	1400
		C32B	13.94	1200				T33B	2.55	592
		C33B	5.56	704				T43B	2.01	337
		C41B	9.43	839				T53B	1.56	444
		C42B	2.92	400				T62B	1.86	356
11-Feb	249	C21B	0.57	96		11-Feb	249	T23B	3.75	1460
		C22B	3.68	412				T33B	1.82	379
		C23B	6.03	567				T43B	1.65	185
		C32B	6.00	437				T53B	1.24	138
		C33B	5.78	900				T62B	0.69	221
		C41B	7.36	578						
		C42B	0.98	218						
8-Mar	274	C22B	7.29	740		8-Mar	274	T23B	5.03	1380
		C23B	14.20	1830				T33B	1.77	734
		C24B	1.16	0.6				T43B	2.40	328
		C31B	0.70	19				T53B	0.54	285
		C32B	2.08	247				T62B	1.17	253
		C33B	1.81	704						
		C41B	1.75	337						
		C42B	1.52	469						
		C22Y	0.72	297						

Note:

(R) Red: Shallow wells, (Y) Yellow: Middle wells, (B) Blue: Deep wells
Iodide is expressed as mg/L
d-MTBE is expressed as µg/L

Test of effectiveness of the treatment process on degradation of MTBE is given by:

$$\text{Upper Confidence Limit (UCL)} = \bar{x} + \frac{t_{(n-1),(1-\alpha)}s}{\sqrt{n}}$$

where $\bar{x}$ is the sample mean of SQRT(d-MTBE) in the Test Plot , $t_{n-1,(1-\alpha)}$ is table look-up t value that reflects the degree of confidence desired, $(1-\alpha, \alpha = 0.2)$ with (n-1) degrees of freedom, s is the standard deviation of the sample, and n is the sample size.

Results: Given $\bar{x}$ = 21.12, n = 54, s = 7.64, $t_{53,0.80}$ = 0.85 for SQRT(d-MTBE) in the Test plot, the UCL is 22.0 which is equivalent to a d-MTBE concentration of 484 μg/L which is greater than the Total Target Level (TTL) for d-MTBE of 5 μg/L.

The test statistic for the one-sided hypothesis test, H_0: $\mu_{SQRT(d-MTBE)} < \mu_0$ (H_a: $\mu_{SQRT(d-MTBE)} > \mu_0$) is given by:

$$t = \frac{\bar{x} - \mu_0}{s/\sqrt{n}}$$

here μ_0 is the square-root of the Total Target Level (TTL) for MTBE which is 5 μg/L. The calculated "t" value is 18.2 and is greater than $t_{54,0.80}$ = 0.84. Therefore the H_0 is rejected: that is, the statement that the mean MTBE concentration in the Test Plot is smaller than 5 μg/L is not true and it is concluded that mean MTBE concentration in the Test Plot is greater than 5 μg/L.

An independent t-test on the square root of d-MTBE concentrations in the Control Plot and in the Test was also run giving the following results:

```
INDEPENDENT SAMPLES T_TEST ON      SMTBE      GROUPED BY    TCCODE

       GROUP                N      MEAN             SD
       0.000 (Control)     93      20.50           8.26
       1.000 (Test Plot)   54      21.11           7.63

SEPARATE VARIANCES T =      -0.452 DF = 118.6 PROB =       0.652
   POOLED VARIANCES T =     -0.443 DF =   143 PROB =       0.659
```

The test indicates that the difference between the Control Plot and the Test Plot is not significant; therefore, the treatment process did not remediate d-MTBE.

Although the use of an ANOVA test to evaluate Test and Control Plots differences was described in the PQA, there are a number of reasons as to why this analysis is not recommended for inclusion in the ITER including:

- Due to the fact that not all monitoring wells provided "useable" data for all sampling events, this becomes a very complicated analysis if performed correctly. One reason for this is that two estimates of experimental error are required, one for the plot type (between monitoring wells) and one for the event (within monitoring wells).

- It is obvious from Figure 7-5 that any event effect has been swamped by variability.

- As shown in Figures 6-1 and 6-2 and Table 7-1, the initial concentrations of MTBE in the Test and Control plots are different.

Discussion

Figure 7-5 shows the time trends of d-MTBE of the ground water in the Test Plot and the Control Plot. The means (solid symbols) of d-MTBE in each sampling event (Event 4 to Event 15) with standard error (error bar) are presented in Figure 7-5. It indicates that the standard errors are so great that differences of d-MTBE between from the Test Plot and the Control Plot are not statistically significant. The average d-MTBE concentration in any sampling event is far greater than the TTL of 5 μg/L.

An alternative way to evaluate if biodegradation of d-MTBE is occurring in the Test Plot is to examine the time trend of total d-MTBE mass with time. Since the d-MTBE front had past the monitoring zone (beyond the line of monitoring wells T61B and T62B in the Test Plot, and C41B and C42B in the Control Plot), time trend of total d-MTBE is not available. However, the time trends of d-MTBE in the transect T23B, T33B, T43B, T53B in the Test Plot, and the transect of T62B, as well as C22B, C32B, and C41B in the Control Plot (Figures 7-6) show that d-MTBE concentrations are erratic and there is no obvious decrease in d-MTBE (SYSTAT, 1990; Keppel, 1982).

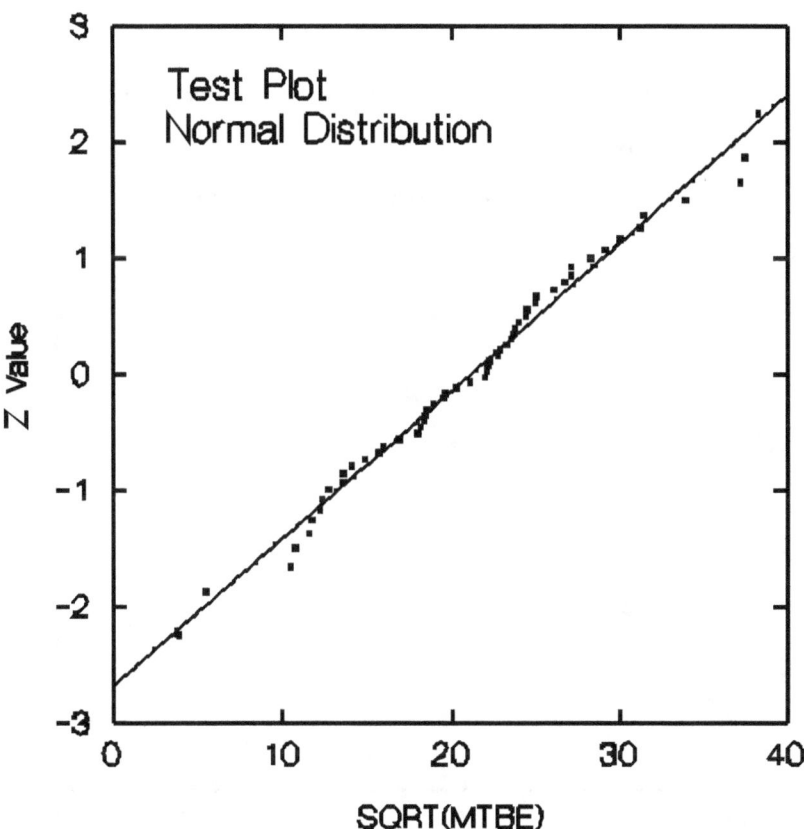

Figure 7-1. Test Plot Normal Distribution

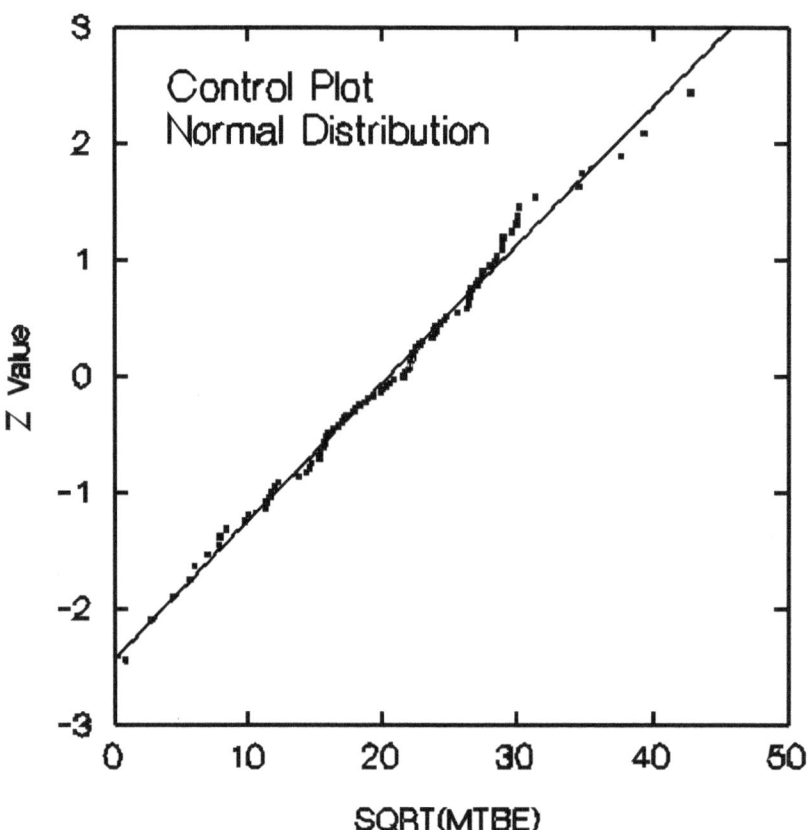

Figure 7-2. Control Plot Normal Distribution

80

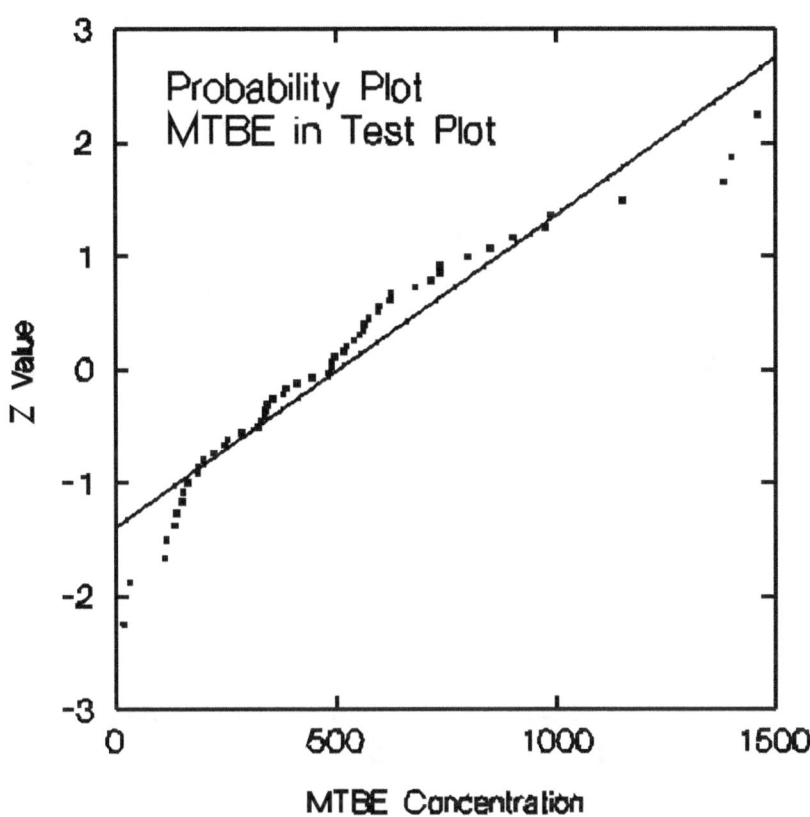

Figure 7-3. Probability Plot MTBE in Test Plot

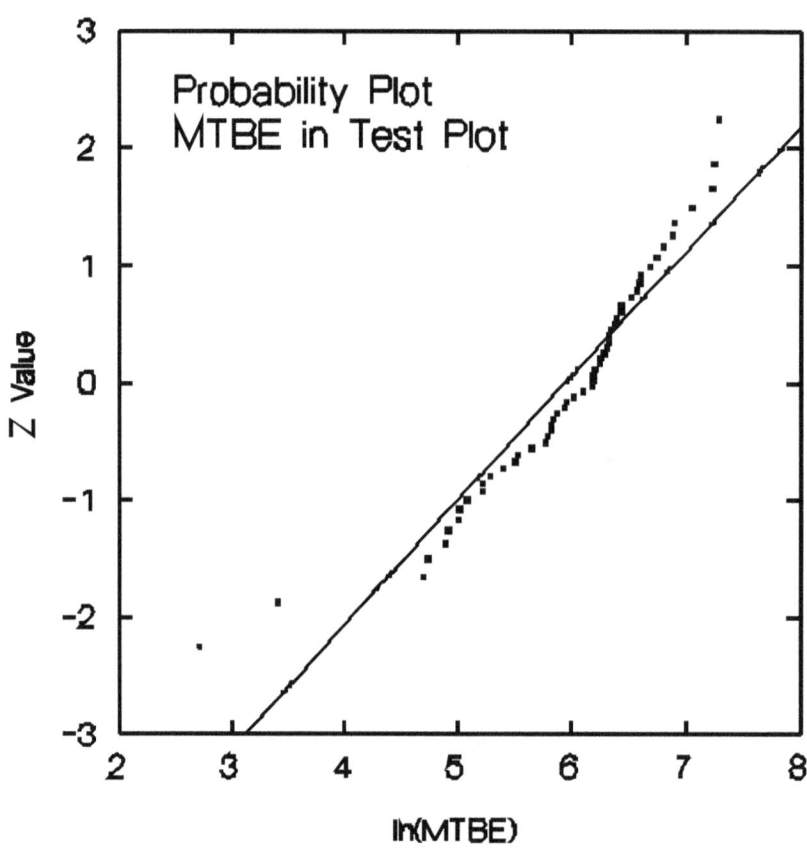

Figure 7-4. Probability Plot MTBE in Test Plot

Time Trends of MTBE for Test and Control Plots

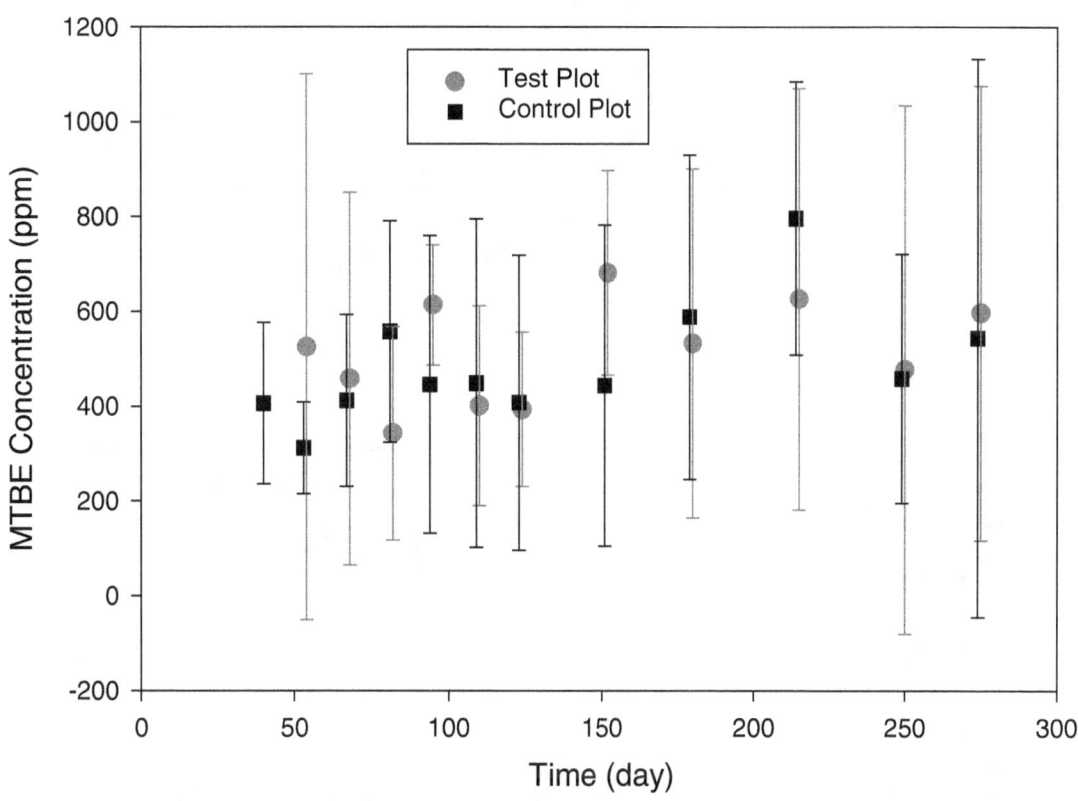

Figure 7-5. MTBE Time Trends for Test and Control plots

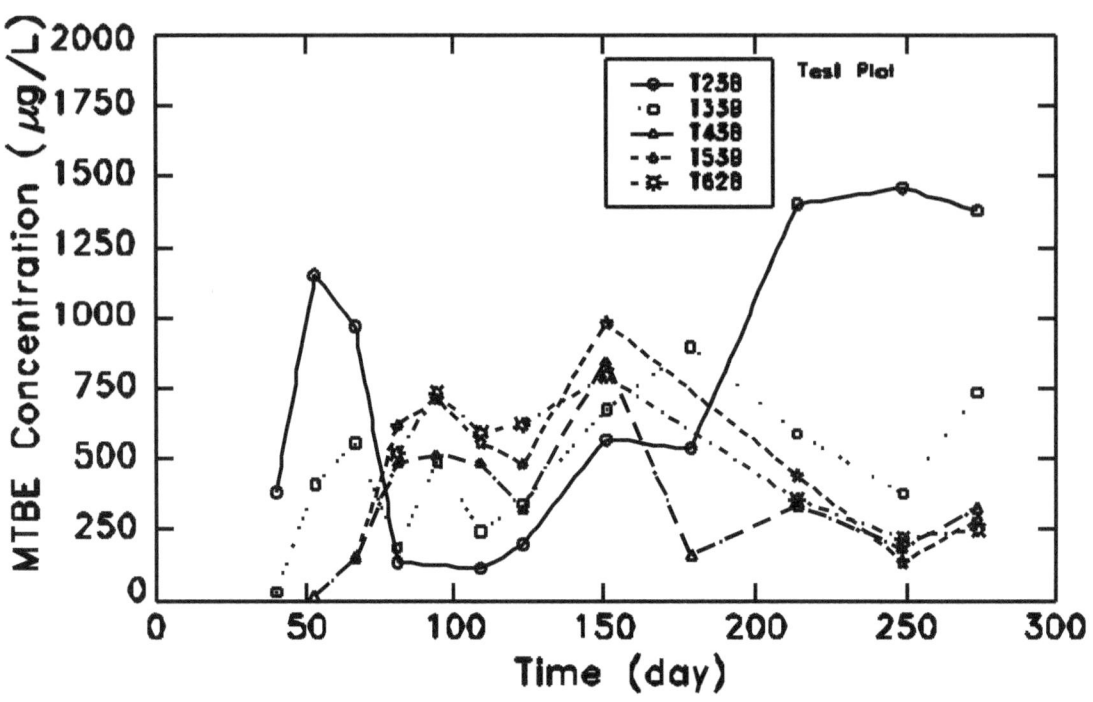

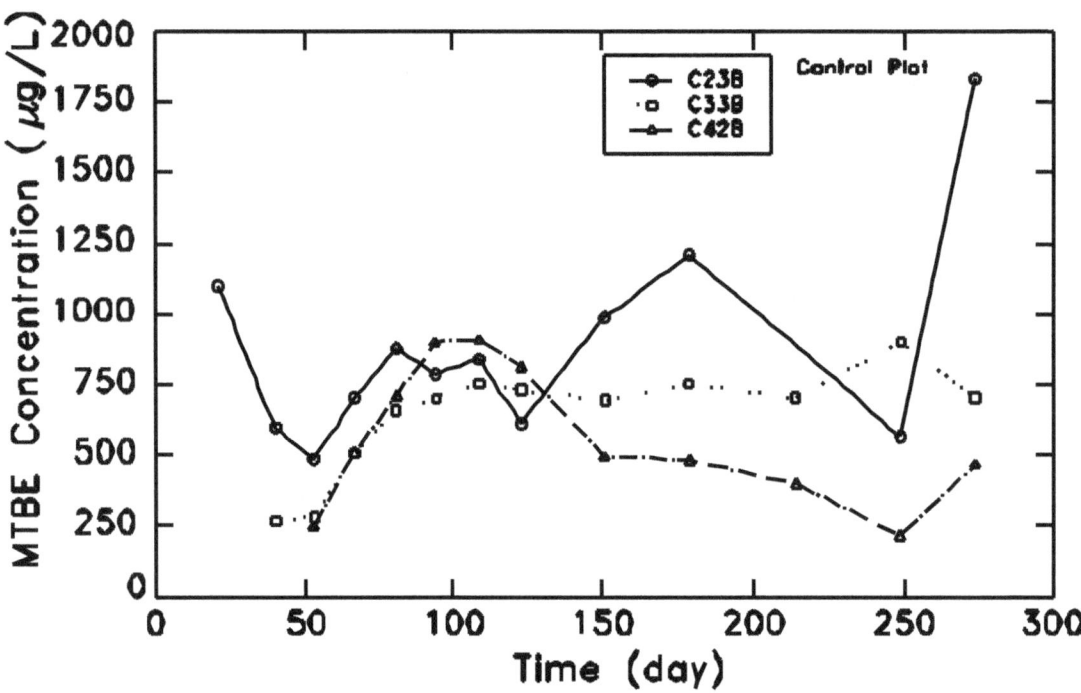

Figure 7-6. Time Trends of Total d-MTBE Mass in Test and Control Plots

84

7.3   EVALUATION OF RESULTS AGAINST OBJECTIVES

This section assesses the results of the Envirogen oxygen and propane biostimulation and bioaugmentation demonstration in relation to stated primary and secondary objectives.

Primary Objective: Will the technology reduce the final levels of MTBE to less than the treatment goals established for the demonstration program?

The primary objective was addressed by measuring d-MTBE concentrations in the "qualified samples" and determining whether, with 80% confidence, the estimate of the population mean is at or below 5 μg/L.

As demonstrated by the t-test in Section 7.2.2, the hypothesis that the mean d-MTBE concentration in the downgradient Test Plot is smaller than 5 μg/L is not true and it is concluded that the mean d-MTBE concentration in the downgradient Test Plot is greater than 5 μg/L.

Because the treatment goal was not achieved for d-MTBE, the primary objective for d-MTBE was not met. Furthermore, in achieving the primary objective, the development of the demonstration evaluation plan centered on providing both direct and indirect means of evaluating the technology to increase support for the study's conclusions. To this end the project plan called for measurements of direct indicators including intrinsic MTBE and introduced d-MTBE, as well as indirect approaches including the creation of daughter products and changes in geochemical parameters.

As discussed previously (Section 6.2.1), the major parameter for evaluating the effectiveness of the technology is by the reduction of intrinsic MTBE. It is important to note that the concentration of MTBE dropped about 0.5 mg/L in October and early November, 2000, and the average of the lower screens in the first downgradient transect in the Test Plot (T21B, T22B, and T23B) dropped from over 5,000 μg/L in September 2000, to 935 μg/L by the time of the first sampling event on June 14, 2001 (Section 6.1). It is also significant, as shown in Figure 6-3 that, although of higher MTBE concentration, the average of the bottom screens in the upgradient injection wells and downgradient Control Plot dropped consistently throughout the period of the demonstration.

The primary goal of the demonstration was to reduce intrinsic MTBE concentrations in the downgradient monitoring wells in the Test Plot to 5 µg/L or below. During the planning stages of the project it was expected that this reduction would necessarily result from intrinsic MTBE levels which, at that time, were 4-5 mg/L. At the beginning of the project, however, as mentioned above, the first transect had already fallen to below 1 mg/L. The intrinsic MTBE in the bottom screens in the downgradient Test Plot for the test period from June 14, 2001, through March 8, 2002, are shown in Figure 7-7 and a least squares line of the data remains between 500 – 600 µg/L (300-800 µg/L actual values) with a small positive slope as determined by a least squares calculation.

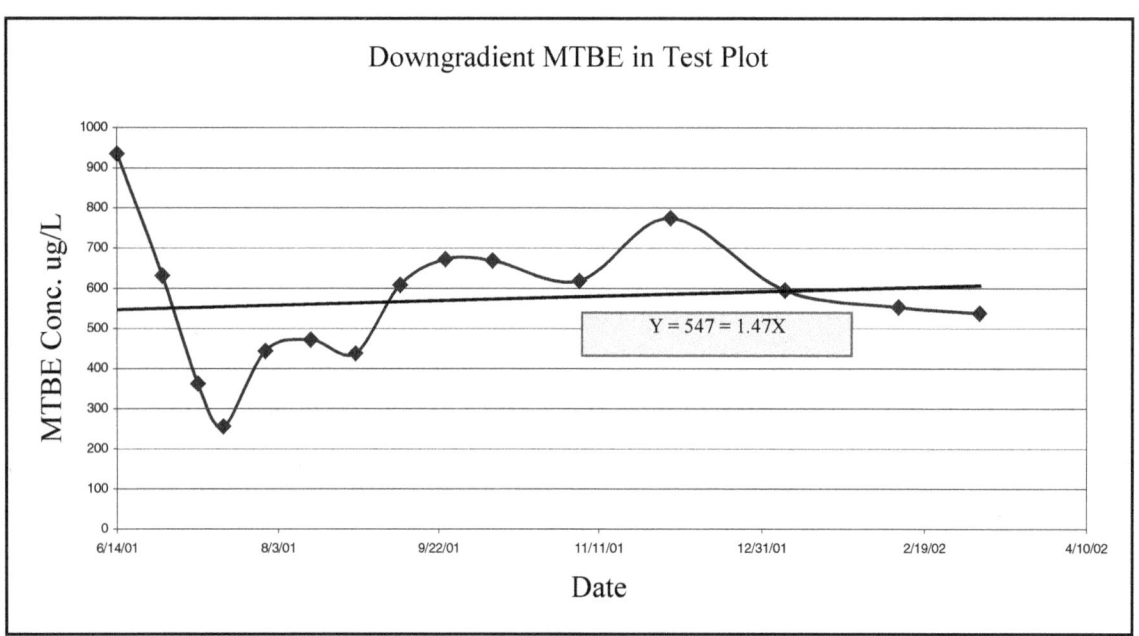

Figure 7-7. Downgradient Test Plot MTBE Concentrations at the Bottom Screens

As discussed in Section 6.2.2, d-MTBE was added as a non-conservative tracer, used in part to avoid possible problems associated with fluctuations in the intrinsic MTBE concentrations. It was introduced through the tracer injection wells in an amount to add about 1.0 mg/L to the downgradient aquifer. Its use would provide direct evidence of the efficacy of remediation by the treatment technology.

As demonstrated by other tracers, its detection varied according to the diverse conductivity paths in both the Test and Control Plots. In general, the more conductive downgradient bottom-screen monitoring wells appeared to increase over the tenure of the project with pronounced variations most likely resulting from periods when the injection wells were not operating, as discussed in Section 6.2.2.

The most important result is demonstrated in Figure 6.8 which shows that d-MTBE concentrations in the lower screens in both the Test and Control Plots increased throughout the period of the evaluation demonstration. Although concentrations in the downgradient Control Plot were somewhat higher than those in the Test Plot, due to zones of higher hydraulic conductivity, it should be noted that the rates of increase, as shown by the least squares line, are the same.

As discussed in Section 6.2.3, the processes resulting in the remediation of intrinsic MTBE and d-MTBE would necessarily result in the production of both deuterated and non-deuterated degradation or daughter products with masses commensurate with the reduction of the parent compounds. Tables 6.2 and 6.3 in Section 6.2.3 show that, as would be expected, only non-deuterated TBA was detected in the bottom screens of both the upgradient Control and Test Plots with mean values of 105 and 29 µg/L, respectively. Non-deuterated TBA also was detected in the downgradient bottom screens with mean values of 63 and 29 µg/L, respectively. Almost identical values of d-TBA were also detected at the downgradient Control and Test Plot with very low mean values of 25 and 29 µg/L, respectively. One other non-deuterated daughter was detected in the downgradient Test Plot, that being acetone with a mean value of 10 µg/L.

Secondary Objective No. 1: Determine time of travel to the sampling points using bromide:

The objective was met as discussed in Section 5.1. Ground-water velocities and associated times of travel were highly variable in both the horizontal and vertical direction. Although the time of travel in the Control Plot was found to be significantly lower than that in the Test Plot, considerable variance was noted even along common transects. A significant finding was that ground-water flow in both Plots was almost totally confined to the lower part of the aquifer. Although highly variable in magnitude and location, ground-water velocities in many locations ranged from 0.1-0.5 ft/day in the more conductive zones.

Secondary Objective No. 2: Establish the absence of trace metals inhibitors:

The objective was met as the analytical results determined from ground-water samples collected from the upgradient monitoring locations during the first sampling event confirmed the absence of the metal inhibitors.

Secondary Objective No. 3: Evaluate the formation of daughter products and determine if they were consistent with a microbiological transformation process:

As discussed in Section 6.2.3, very low levels of TBA and d-TBA were detected in almost equal amounts in both Control and Test Plot downgradient wells. Low levels of non-deuterated TBA were also detected in the upgradient wells in both the Control and Test Plots.

Section 6.2.4 discusses observations in geochemical parameters as another indirect measure of the effectiveness of the treatment technology. The biological processes involved in the reduction of MTBE or d-MTBE would necessarily result in alterations to the alkalinity, nutrients, and electron donors such as organic carbon. As shown in Tables 6.3 and 6.4, not only are the geochemical parameters very similar between the Control and Test Plots, they remain virtually unchanged from upgradient to downgradient monitoring locations.

Secondary Objective No. 4: Evaluate changes in geochemical parameters and determine if they were consistent with the microbiological transformation processes:

As discussed in Section 6.2.4, water quality parameters such as alkalinity and nutrients remained constant throughout the test period indicating that significant alterations resulting from biological processes were not taking place. Of particular significance is the lack of utilization of organic carbon which is required by bacteria as electron donors in considerably larger mass quantities than the mass of the contaminant being remediated.

Secondary Objective No. 5: Define operating costs over a 10-month period of stable operation:

Operating costs are discussed in the economic analysis of the Envirogen technology in Section 8. The economic analysis utilized operating data from a previous demonstration as well as information collected from field demonstrations at hazardous waste sites.

Secondary Objective No. 6: Estimate exponential order of degradation and calculate MTBE degradation rate constant:

According to the statistical analysis Section 7.2.2, an alternative way to evaluate if biodegradation of d-MTBE is occurring in the Test Plot is to examine the time trend of total d-MTBE mass with time. Since the d-MTBE front had passed the monitoring zone (beyond the line of monitoring wells T61B and T62B in the Test Plot, and C41B and C42B in the Control Plot), time trend of total d-MTBE is not available. However, the time trends of d-MTBE in the transect T23B, T33B, T43B, T53B in the Test Plot, and the transect of T62B, as well as C22B, C32B, and C41B in the Control Plot (Figure 7.6) show that d-MTBE concentrations are erratic and that there is no obvious decrease in d-MTBE.

Secondary Objective No. 7: Determine the fraction of d-MTBE removed at each sampling location at each sample time:

As discussed in Section 6, the most persuasive evidence of this finding is shown in Figure 6.8 which demonstrates that at downgradient bottom screens d-MTBE in both the Control and Test Plots increased throughout the test period. Therefore there was no significant reduction of d-MTBE even though minor levels (See Tables 6-2 and 6-3) of d-TBA at about the same concentration were detected in both the Test and Control Plots.

Secondary Objective No. 8: Evaluate d-MTBE reduction in the Control Plot receiving only oxygen injection:

As shown in Figure 6.8, the least squares fit in both the Control and Test Plots indicates that downgradient d-MTBE concentrations increased over the study period at about the same rate, with those in the Control Plot being somewhat higher.

7.4   QUALITY ASSURANCE AND QUALITY CONTROL RESULTS

A data quality review was conducted by SPRD to evaluate field and laboratory QC results, the implications of QC on the overall data quality, document data use limitations for data users, and remove unusable values from the demonstration data sets. The results of this review were used to produce the final data sets to assess the treatment technology and to draw conclusions. The QC data were evaluated with respect to the quality assurance (QA) objectives defined in the project QAPP (SAIC 2001).

The analytical data for ground-water samples collected during the Envirogen demonstration were reviewed to ensure that they are scientifically valid, defensible, and comparable. A data quality review was conducted using both field and laboratory QC samples. The field QC samples included source water blanks, field blanks, trip blanks, matrix spike/matrix spike duplicates (MS/MSD), and sample duplicates. Laboratory QC checks included laboratory blanks, surrogate spikes, and laboratory control sample/laboratory control sample duplicates (LCS/LCSD) (also known as blank spike/blank spike duplicates). Initial and continuing calibration results were also reviewed to assure the quality of the data and that proper procedures were used. The review focused on assessing the precision, accuracy, completeness, representativeness, and comparability of the data.

All critical parameter data were reviewed with one hundred percent of the iodide samples by SPRD and at least ten percent of d-MTBE samples by SAIC from the demonstration phase being fully validated (recalculated from the raw instrument data). In addition to the above QC checks, reviews of sample chain of custody, holding times, and critical parameter identification and quantification were performed by SPRD.

Overall, the data quality review assessed the critical parameter data to be usable for the purpose of evaluating the technology and the attainment of the primary objective for this demonstration. In some instances, results for one or more QC parameters were outside of control limits; however, deviations were generally slight, and no broad qualifications of data or other actions were required. A description of the more significant deviations from QC acceptance criteria and the limited impact of these deviations are described below:

- During the fifteen sampling events iodide was not detected in any of the trip blanks. Another critical parameter, d-MTBE, was only detected in one trip blank during sampling Event 4 at 0.1 µg/L. Because this is an order of magnitude below the reporting limit, it does not have any impact on data quality.

- The laboratory (ALSI) that performed the MTBE and d-MTBE analysis routinely takes water for blanks and standards from an on-site well since the well-water contains very few detectable VOCs. This water does contain background concentrations of MTBE at about 0.3 µg/L. MTBE was detected at concentrations of approximately 0.3 µg/L in many of the method blanks and in 13 of the trip blanks. Because this value is approximately 3 times less than the reporting limit for MTBE for this project (and approximately 16 times less than the treatment goal), it was determined very early on in the project that this water could be used without affecting project activities.

- For the VOC analyses, during Events 1 - 15, MS and MSD percent recoveries were generally (80/84 for MTBE/d-MTBE, 84/84 for 2-propanol, 84/84 for acetone, and 84/84 for TBA) within the acceptance criteria of 80 to 120 percent, and no data were rendered unusable due to MS/MSD results. In some cases (4/84 for MTBE/d-MTBE), the percent recoveries for MTBE and other critical parameters were above the QC limits in MS/MSDs performed on water samples. However, these high recoveries were most likely due to the high native concentrations present in the sample (i.e., the spike concentration was too low) or in the case of iodide, it was due to using a small volume for spiking (again, the spike concentration was too low). Because the LCS/LCSD recoveries were generally (62/62 for MTBE/d-MTBE, 62/62 for 2-propanol, 62/62 for acetone, and 61/62 for TBA) within the acceptance criteria, data were not disqualified based on the high MS/MSD recoveries. Relative percent differences (RPDs) between the MS and MSD samples were also generally (39/42 for MTBE) within the acceptance limits.

- For VOC, LCS/LCSD percent recoveries and RPDs were generally (62/62 for MTBE) within QAPP acceptance limits. Similar accuracy was observed from the recoveries of the VOCs surrogates from the demonstration samples.

- Field duplicates were collected and analyzed at a frequency of five percent or more for the fifteen demonstration sampling events. Field duplicate results uniformly met QAPP precision criteria of +/- 25% RPD for the critical parameters. Therefore, no qualification was added to the data.

SPRD also conducted a cursory quality control review for the conservative tracers used as non-critical analytical parameters. This review was performed to confirm the overall usability of the data in the evaluation of the secondary objectives. Based on this review, the non-critical data were assessed to be usable for their intended uses.

During the third demonstration sampling event, QA supervisory personnel conducted a Technical Systems Audit (TSA) of field sample collection and handling procedures. In order to verify that the requirements of the EPA QAPP were met, QA supervisory personnel also completed two TSAs of the laboratories responsible for analyzing the critical parameters (d-MTBE/MTBE and iodide). Given that the three TSAs were conducted early on in the project, the non-conformance had minimal impact on data quality.

Based on the information reviewed during the field TSA and requirements of the QAPP, four findings and three observations were noted. The field TSA also resulted in clarifications and modifications to the sampling procedures established in the QAPP. These generally involved changes in documentation practices, sampling order schedules, sample packing procedures, and sample identification number formats. In addition, the field TSA corrective action included documentation of adequate field observations so that various events could be reconstructed. The audit increased the frequency of operation and maintenance activities performed by the NBVC personnel to a minimum of five times per week. A requirement to ensure that the correct concentration of d-MTBE and iodide is present in the tracer reservoirs (Teldar bags) resulted in monthly sampling throughout the duration of the project.

The laboratory audit of ALSI for the analysis of d-MTBE/MTBE identified no finding or observation. The only identified miscellaneous issues dealt with an on-site well that was used as the source water for blanks and preparation of aqueous standards as described above. The audit resulted in minor modifications to the QAPP involving personnel name and title corrections as well as the addition of new personnel responsible for COC issues.

For the analysis of iodide samples, it is paramount to note the circumstances surrounding the involvement of ManTech in performing this task. Since, initially, DelMar was retained to conduct the analysis, the samples from the first sampling event (6/14/2001) were submitted and analyzed by DelMar. Prior to the second sampling event on 6/20/2001, due to budgetary constraints, a decision was made to send all subsequent samples to ManTech. Samples from the second sampling event on 6/28/2001, and all subsequent events were submitted to ManTech. Therefore, ManTech had a very short lead time to prepare to perform these analyses, which resulted in the following audit QAPP discrepancies.

During the ManTech TSA, while the third of fifteen sampling events was taking place, the following procedures were audited: sample receipt and storage; iodide analysis; reporting, reduction, and validation of data; and requirements of the QAPP. Four findings and six observations were identified. The findings concerned the addition of the second source standard for iodide which was analyzed and was within acceptance criteria but was not reported for four of the sampling events. Additional TSA recommendations were to also perform LCS analysis for every sampling event with each MS/MSD, to accumulate all data storage locations with respect to this evaluation project in one location and to identify this location on Millennium Software. The TSA increased the frequency of the routinely performed raw data backups on CDs with their inferences being referenced in the Lab notebook.

Therefore, as corrective actions, in response to the above noted findings, ManTech prepared an addendum to the respective report letters for Events 2,3,6, and 7 that contained the results for the second source standards. All other future reports included the second source standard values. Laboratory Control Samples (LCS) was analyzed for all the events that followed the audit debriefing and all subsequent events with each MS/MSD. Since the data acquisition computer is equipped with a CD writer, raw data backups were performed after analyzing each sample event. All backups and data locations were identified in the Lab notebook.

The last finding of the laboratory audit described the spike recovery of the fourth sampling event. The report stated that MS/MSD recoveries on sample 4-T16d for iodide were incorrectly calculated due to improper integration performed by the software and had to be done manually. Following the manual corrections, MS recovery at this location (4-T16d) was at 129% which exceeded the QAPP acceptance criteria of 80-120%. This high recovery is probably due to the chemist using a small volume (10 μL) for spiking because small volume measurements are subject to greater errors since all other QC during the fourth sampling event met criteria. Consequently, it was determined that this non-conformance did not negatively affect data quality and all data were used without qualification. Also a change in the true value of the check standard from 10 ppm to 1 ppm was realized. Therefore, as a corrective action, the laboratory issued an amendment to the report letter for Event 4 which included the corrected values for the spike recoveries on sample 4-T16d. The final concentration of spike added was 1.0 mg/L. All values for the check standards that were incorrectly reported were also changed.

Three of the observations noted during the ManTech laboratory TSA were (1) to provide a temperature log book with daily temperature recordings for the sample storage room, (2) a need to issue a report to the TLP to identify all analytical conditions for the IC method used (based on EPA Method 300.0) with detailed descriptions to be able to reproduce the method, and (3) a recommendation to perform a calibration check for autopipets used for spiking according to a draft SOP developed by the SPRD QA. This corrective action was implemented following the audit debriefing and indicated the autopipets calibration fell within the required criteria.

The ManTech TSA recognized a fourth observation that the MDL had not been determined for iodide. The PQL was defined as 0.5 mg/L in the QAPP. Accordingly, all iodide data for samples with quantities between 0.5 mg/L and the MDL were required to be reported (with qualifiers). Consequently, all sample data previously reported as <0.5 mg/L were reviewed to determine if they fell in this range, and reports were re-issued with these data. The TSA report stated that the non-conformance should have minimal impact on the data quality for the primary objectives.

The only minor clarification and modification to the QAPP resulted from change of a personnel name to correctly identify the sample recipient personnel and duty performed. This change did not negatively affect data quality.

SECTION 8
ECONOMIC ANALYSIS

This economic analysis presents cost estimates for using the Envirogen technology to treat contaminated ground water. Cost data were compiled during the demonstration at the NBVC, during the previous demonstration at a service station in New Jersey, and from Envirogen. The vendor claims that because the demonstration at the NBVC project involved the application of their technology at pilot-scale, it was not possible to evaluate the start-up costs based on data collected during the demonstration. Therefore, the start-up costs were reviewed and scaled accordingly based on team member's experience with full-scale remediation using related technologies (i.e., sparging and biostimulation) and lessons learned during the demonstration. This also is due, in part, because some of the demonstration's associated costs such as installation of the ground-water monitoring wells and plots survey were handled by EPA. As a National Environmental Technology Test Site (NETTS), NBVC was responsible for providing access to pre-characterization data. Furthermore, the ground-water sampling and routine operation and maintenance of the Envirogen system was performed in association with NBVC staff. NETTS supplied utilities, handled waste disposal, and assisted in site demobilization activities. EPA served as the interface between California Water Quality Board and the NBVC for the technical justification and preparation of the project permit, therefore, the permitting cost for Envirogen was reduced to attending a public hearing meeting.

This section describes a site, based on experience gained by Envirogen's previous demonstration at a gas station as well as that gained at the Port Hueneme field demonstration. These experiences were selected for economic analysis, summarized the major issues involved and assumptions made in performing the analysis, discussed costs associated with using the Envirogen propane biostimulation and bioaugmentation technology to treat ground-water contaminated with MTBE, and presented a conclusions of the economic analysis.

8.1 INTRODUCTION

The vendor operated a system consisting of a network of oxygen, bacteria, and propane injection points, pressurized oxygen and propane gas delivery and control systems, and ground-water and soil-gas monitoring network. However, the vendor claims the system could operate with slight modifications at a larger or smaller scale; therefore, the economic analysis presents and evaluates costs based on an application involving the treatment of contaminated ground water at a typical gas station site. Table 8-1 summarizes estimated costs as determined by Envirogen.

8.2   APPLICATION ISSUES AND ASSUMPTIONS

Typically, costs are placed in 12 categories applicable to typical cleanup activities at Superfund and RCRA sites (Evans 1990). These categories include: (1) site preparation, (2) permitting and regulatory, (3) mobilization and startup, (4) equipment, (5) labor, (6) supplies, (7) utilities, (8) effluent treatment and disposal, (9) residual waste shipping and handling, (10) analytical services, (11) equipment maintenance, and (12) site demobilization. Even with a detailed analysis, costs are considered to be order-of-magnitude estimates with an expected accuracy of from 30 - 50 percent above to 30 -50 percent below actual costs. Therefore, for this economic analysis, the categories applicable to hazardous waste sites are recognized and discussed.   In the event that a determination of a distinct cost associated for each of the categories was not possible due to the special circumstances of this project (see Section 8), an attempt was made to provide an estimated cost at the hazardous waste sites. However, based on Envirogen's past performance, this section also describes the case of "a typical gas station" selected for economic analysis, summarizes the major issues involved and assumptions made in performing the analysis, discusses costs associated with using the Envirogen technology to treat ground-water contaminated with MTBE, and presents the conclusions of the economic analysis.

This section lists the major assumptions, site-specific factors, equipment and operating parameters, and financial calculations used in this economic analysis of the Envirogen technology. Issues and assumptions are presented in Sections 8.2.1 through 8.2.3. Certain assumptions were made to account for variable site and waste parameters. Other assumptions were made to simplify cost estimating for situations that actually would require complex engineering or financial functions. Section 8.2.3 provides a hypothetical base-case scenario developed from the assumptions. In general, Envirogen system operating issues and assumptions are based on information provided by Envirogen and observations made during the demonstration.

TABLE 8-1
Estimated Cost for Envirogen Propane Biostimulation and Bioaugmentation Project at a Typical Gas Station

Activity	Event No.	Labor	Pass Through	Subcontracted Equipment	Materials	Total
Design	1	$ 21,700	$ -	$ -	$ -	$ 21,700
Procurement and Mobilization	1	$ 19,540	$ 120	$ 2,625	$ -	$ 22,285
Installation	1	$ 31,660	$ 1,350	$ 15,015	$ 1,815	$ 49,840
Baseline Monitoring	1	$ 1,400	$ 1,550	$ 1,208	$ -	$ 4,158
Startup	1	$ 4,360	$ 480	$ -	$ -	$ 4,840
O&M and Quarterly Monitoring	8	$ 6,135	$ 2,005	$ 53	$ 28	$ 65,760
Utilities Including Electric and Propane/ Quarter	8		$ 430			$ 3,440
Demobilization	1	$ 3,325	$ 300	$ -	$ -	$ 3,625
Final Report	1	$ 1,605	$ -	$ -	$ -	$ 1,605
Total*						$ 177,253

Abbreviation:

O&M: Operation and Maintenance

Note:

*:   Total estimate for remediation cost is based upon 2 years of operation.
1.   The cost of oxygen is not provided in Table 8-1.
2.   Design cost includes design and drawings, discharge permit application, and attending one meeting.
3.   Procurement and mobilization include equipment and materials procurement, mobilization preparation, and mobilization.
4.   Installation cost includes subcontractors' labor, materials, and equipment for site work including air sparging points, monitoring wells, trenching, and pipe installation, backfilling and surface restoration, system and electrical connection.
5.   Baselines monitoring includes sampling 4 wells and VOC analysis.
6.   Startup cost is based on three days of monitoring and a letter report.
7.   Quarterly monitoring includes sampling 4 wells and VOC analysis and a letter report.
8.   Demobilization includes disconnection, dismantling, and system removal from site.
9.   Final report includes final letter report prepared and submitted to client.

8.2.1 Site-Specific Factors

Site-specific factors can affect the costs of using the Envirogen treatment system. These factors can be divided into the following two categories: waste-related factors and site features. Waste-related factors affecting costs include waste volume, contaminant types and levels, treatment goals, and regulatory requirements. Waste volumes affect total project duration and, consequently, costs because a larger volume takes longer to treat. However, economies of scale are realized with a larger-volume project when the fixed costs are distributed over the larger volume. The contaminant types and levels in the ground water and the treatment goals for the site determine (1) the appropriate Envirogen treatment system size, which affects capital equipment costs , and (2) periodic sampling requirements, which affect analytical costs. Regulatory requirements affect permitting costs and sampling as well as the ground-water monitoring costs. Site features affecting costs include ground-water recharge rates, ground-water chemistry, site accessibility, availability of utilities, and geographic location. Ground-water recharge rates affect the time required for cleanup. Site accessibility, availability of utilities, and site location and size all affect site preparation costs. Site-specific assumptions include the following:

1. The site is a located near an urban area. As a result, utilities and other infrastructure features (for example, access roads to the site) are readily available.

2. The site is located in a region that has relatively mild temperatures during the winter months resulting in potentially high bacterial metabolism.

3. Contaminated ground water is located in a shallow aquifer.

8.2.2 Equipment and Operating Parameters

The Envirogen biostimulation system can be used to treat shallow ground water contaminated with MTBE. This analysis provides costs for treating contaminated ground water. Envirogen will provide the appropriate system configuration based on site specific conditions, of which ground-water recharge rates and contaminant concentration are the primary considerations. The Envirogen system can be configured to meet certain site requirements by varying the sparge systems, which are also dependent on site conditions. The Envirogen system is modular in design, which allows for treatment units either in series or in parallel to treat ground water. This analysis focuses on the estimated costs associated with the unit demonstrated at the NBVC Site. The vendor claims that their system can treat ground water contaminated with BTEX/MTBE concentration in the source area at 60 mg/L with the maximum contaminant being MTBE. The system operates on a continuous cycle, 24 hours per day, 7 days per week.

Based on these assumptions, this analysis assumes that about 81,000 gallons of water need to be treated to complete the ground-water remediation project, which will take about 2 years to process. It is difficult in practice to determine both the volume of ground water to treat and the actual duration of a project, but these figures have been assumed to perform this economic analysis.

As expected in a full operation, neither depreciation nor salvage value is applied to the costs presented in this analysis because the equipment is not purchased by a customer. All depreciation and salvage value is assumed to be incurred by Envirogen and is reflected in the ultimate cost. Equipment and operating parameter assumptions are listed below.

1. The treatment system is operated 24 hours per day, 7 days per week, 52 weeks per year;

2. The treatment system operates automatically without constant attention of an operator;

3. Modular components consisting of the equipment needed to meet potential treatment goals are mobilized to the site and assembled by Envirogen;

4. Air emissions monitoring is necessary; and

5. Envirogen equipment will be maintained by Envirogen and will last for the duration of the ground-water treatment project with proper maintenance.

Specifically, Envirogen claims that operation and maintenance costs shown in Table 8-1 are based on typical monitoring requirements including:

1. Personnel training required to operate, maintain, and monitor the system;

2. Analytical costs;

3. Routine maintenance;

4. Waste handling and disposal; and

5. Utilities.

Envirogen believes that no specialized training costs are associated with the operation, maintenance, and monitoring of this type of system. An understanding of system operation and the importance of vapor monitoring results as they apply to fugitive VOC and propane emission is required. Analytical costs for MTBE analysis would not increase for the typical gas station site at which regular VOC analysis is constructed, as MTBE is included in the standard VOC scan. Additional analytical costs might include analysis for TBA, dissolved carbon dioxide, and propane. Bacterial analyses may be required at some sites, with associated additional costs, particularly at sites where bioaugmentation is performed. Routine system maintenance, including that necessary to prevent silting and clogging of wells, is similar to that required for a typical air sparge system at a comparable cost. The labor costs for sampling and monitoring activities would be slightly higher than those for a standard monitoring program, because low-flow ground-water sampling methods would be employed.

8.2.3   Base-Case Scenario

A hypothetical base-case scenario has been developed using the issues and assumptions described above for the purposes of formulating this economic analysis. Although the system under this evaluation was not a portable unit, the costs presented are for an Envirogen system for the remediation of contaminated ground water at a typical gas station. Thus, the following assumptions are made by Envirogen for the gas station remediation.

1. The service station area is 100 feet. x 60 feet. with the remediation area measuring 60 x 60 ft.

2. The subsurface soil is a medium sand with a porosity of 0.3 and the depth to ground water is 10 ft. below grade (bg).

3. The vertical extent of ground water contamination is 10 feet. below the ground water. Thus, the volume of ground water to be treated is 81,000 gal. The volume of saturated contaminated soil is 1330 yd^3.

4. The BTEX/MTBE concentration in the ground water in the source area is 60 ppm with the maximum contaminant being MTBE.

Envirogen made additional assumptions for the installation, operation, and maintenance of their biostimulation system:

1. 6 air sparging/propane injection points installed to 10 feet. below ground water.

2. 4 monitoring wells installed to 10 ft. below ground water.

3. 4 monitoring points installed to 1 foot. above ground water.

4. Estimated 200 feet. of piping to injection points installed below grade.

5.   Biostimulation system trailer with air sparging blower, propane tank, piping, instrumentation and control panel. The tasks for implementing the design, installation, and operation and maintenance of the system with a description of the subtasks with their associated costs are provided in Table 8-1.

Envirogen claims that the total cost is based on the time needed to remediate the ground water to a cleanup objective of 70 µg/L. The time to remediate the ground water to the cleanup objective is estimated to be two years which was derived from degradation rates from other sites. Based on a two-year remediation, the total cost for the project is estimated to be $177,000 +/- 20%. Envirogen stated that at a volume of contaminated ground water of 81,000 gallons, the unit cost to remediate this medium is $2.35/gal. Further assumptions used for this base-case scenario are listed below.

- The air sparging will operate four times a day at 0.5 hour each time for a total operating time of 2 hours/day.

- The site is near Envirogen's office and travel cost and per diems are not needed.

- If bacterial injection is needed, the additional cost is $1000 per event.

- The biostimulation system will be leased to the project.

8.3   COST CATEGORIES

Table 8-1 presents cost breakdowns as provided by Envirogen addressing the various cost categories. Cost data associated with the MTBE demonstration program and hazardous waste sites have been presented for the following categories: (1) site preparation, (2) permitting and regulatory, (3) mobilization and startup, (4) equipment, (5) labor, (6) supplies, (7) utilities, (8) effluent treatment and disposal, (9) residual waste shipping and handling, (10) analytical services, (11) equipment maintenance, and (12) site demobilization. Each of these cost categories is discussed below.

8.3.1   Site Preparation Costs

Site preparation costs include administrative, treatment area preparation, treatability study, and system design costs. Site preparation administrative costs, such as costs for legal searches, access rights, and site planning activities, are usually estimated to be $35,000.

The treatment area preparation includes constructing a shelter building or purchasing a pre-manufactured shed for the housing of the air sparging blower, propane tank, piping, instrumentation and control panel. The shelter building needs to be constructed before mobilization of the technology system.

A building with a minimum of 200-square-foot is required for the system. Vendor will provide the shelter building design specifications. Construction costs will be varied based on the geographic location and the need for installation of heating and cooling system. Construction cost for building a shelter is estimated to be $90 per square foot, with a natural gas heating and cooling unit and ductwork costing about $10,000 installed. The total shelter building construction cost system is estimated to be $28,000.
This analysis assumes that monitoring wells exist on site and are located 200 feet from the shelter building. The total costs, including all electrical equipment and installation (air sparging blower and instrumentation and control panel), are $7,000. Piping and valve connection costs are about $20 per foot, which covers underground installation. Therefore, the total piping costs are $4,000. The total treatment area preparation costs are estimated to be $74,000.

A treatability study and system design will be conducted by the vendor to determine the appropriate treatment system. It is assumed that the vendor will transport its mobile system to the site to test the equipment under site conditions. Six to eight samples will be collected from the influent and effluent and will be analyzed off site for VOCs. The estimated treatability study cost is $15,000, including labor and equipment costs. System design includes determining the size and configuration of the system to achieve treatment goals and designing the configuration. The system design is estimated to cost $5,000. Total site preparation costs are, therefore, estimated to be $94,000.

8.3.2 Permitting and Regulatory Costs

Permitting and regulatory costs depend on whether treatment is performed at a Superfund or a RCRA corrective action site and on how treated water and any solid wastes are disposed. Superfund site remedial actions must be consistent with all applicable environmental laws, ordinances, regulations, and statutes, including federal, state, and local standards and criteria. Remediation at RCRA corrective action sites requires additional monitoring and record keeping, which can increase the base regulatory costs. In general, applicable or relevant and appropriate requirements (ARARs) must be determined on a site-specific basis. The cost of this permit would be based on regulatory agency requirements and treatment goals for a particular site. The discharge permit is estimated to cost $5,000.

8.3.3 Mobilization and Startup Costs

Mobilization and startup costs include the costs of transporting the system to the site, assembling the system, and performing the initial shakedown of the treatment system. The vendor provides trained personnel to assemble and conduct preliminary tests on the system. The vendor personnel are trained in health and safety procedures, so health and safety training costs may not be included as a direct startup cost. Initial operator training is needed to ensure safe, economical, and efficient operation of the system. The vendor provides initial operator training to its clients as part of providing the system. Transportation costs are site-specific and vary depending on the location of the site in relation to the system. For this analysis, the system is assumed to be transported 1,000 miles. The vendor retains the services of a cartage company to transport all of their treatment system equipment. Mobilization costs are about $10 per mile, for a total cost of $10,000. The costs of highway permits for overweight vehicles are included in this total cost. Assembly costs include the costs of unloading equipment from the trailers, assembling the system, hooking up well piping, and hooking up electrical lines. A two-person crew will work three 8-hour days to unload and assemble the system and perform the initial shakedown. The total startup costs are about $10,000, including labor and hookup costs. Total mobilization and startup costs are therefore estimated to be $20,000.

Specifically, for the purpose of this economic analysis, as described previously in the Section 8.0, the startup costs were reviewed and scaled accordingly based on Envirogen team member's experience with full-scale remediation using related technologies (i.e., sparging and biostimulation) and lessons learned during the NBVC demonstration. Each of the costs is site-specific and will vary according to the degree of design and installation required. Startup costs that were evaluated include the following:

1. System design and work plan preparation;

2. Permitting and regulatory approval;

3. Well installation costs including air sparge points and monitoring wells; and

4. Capital equipment costs including system components, and monitoring equipment, and well installation costs are not applicable if an existing system (e.g., an air sparge system) is being retrofitted to include propane injection and bioaugmentation. In that case, existing monitoring wells would be used, and existing air sparge points could be used for substrate and bacterial injection. According to Envirogen, capital equipment costs for system components associated with retrofitting an existing system are minimal.

Envirogen further claims that in any propane biostimulation system, very little propane is required, with typical feed rates of less than 0.3 pounds of propane per day. When coupled with air or oxygen injection, the need for vapor extraction is typically eliminated, although the need for this contingency is site-specific. If a vapor extraction system is required, the cost for a standard SVE system would apply.

8.3.4   Equipment Costs

Envirogen provides the complete Envirogen treatment system configured for site-specific conditions. All Envirogen treatment equipment is leased to the client. As a result, all depreciation and salvage value is incurred by Envirogen and is reflected in the price for leasing the equipment. At the end of a treatment project, Envirogen decontaminates and demobilizes its treatment equipment (see Section 8.3.12, Site Demobilization Costs). Envirogen assumes that this equipment will operate for the duration of the ground-water remediation project and will still function after the remediation is complete as a result of routine maintenance and modifications. Equipment costs are determined by the size of the Envirogen system needed to complete the remediation project and are incurred as a lump sum; as a result, even though the equipment is leased to the client, it is not priced at a monthly rate. For this analysis, Envirogen estimates that the base capital equipment costs is $10,000 for a system employed at a typical gas station.

8.3.5   Labor Costs

Once the system is functioning, it is assumed to operate continuously except during routine maintenance, which the vendor conducts (see Section 8.3.11, Equipment Maintenance Costs). One operator trained by the vendor performs routine equipment monitoring and sampling activities. Under normal operating conditions, an operator is required to monitor the system about once each week. This analysis assumes that the work is conducted by a full-time employee of the site owner and is assigned to be the primary operator to perform system monitoring and sampling duties. Further, it is assumed that a second person, also employed by the site owner, will be trained to act as a backup to the primary operator. Based on observations made at the demonstration, it is estimated that operation of the system requires about 8 hours per week of the primary operator's time. Assuming that the primary operator's burden labor rate is $50 per hour, the total annual labor cost is estimated to be $20,800.

8.3.6   Supply Costs

Except for oxygen, propane, and bacteria, no other chemicals or treatment additives are expected to be needed to treat the ground water using the technology. Supplies that will be needed as part of the overall ground-water remediation project include Level D, disposable personal protective equipment (PPE), PPE disposal drums, and sampling and field analytical supplies. Disposable PPE typically consists of latex inner gloves, nitride outer gloves, and safety glasses. This PPE is needed during periodic sampling activities. Disposable PPE is assumed to cost about $600 per year for the primary operator. Used PPE is assumed to be hazardous and needs to be disposed of in 24-gallon, fiber drums. One drum is assumed to be filled every 2 months, and each drum costs about $12. The total annual drum cost is, therefore, about $100.

Sampling supplies consist of sample bottles and containers, ice, labels, shipping containers, and laboratory forms for off-site analyses. For routine monitoring, laboratory glassware is also needed. The numbers and types of sampling supplies needed are based on the analyses to be performed. Costs for laboratory analyses are presented in Section 8.3.10. The sampling supply costs are estimated to be $1,000 per year. Total annual supply costs are estimated to be $1,700.

8.3.7   Utility Costs

Electricity is the only utility used by the Envirogen system. Electricity is used to run the Envirogen treatment system. This analysis assumes that electrical power lines are available at the site. Electricity costs can vary considerably depending on the geographical location of the site and local utility rates. Also, the consumption of electricity varies depending on the Envirogen system used, the total number of air sparging units and other electrical equipment operating. This analysis assumes a constant rate of electricity consumption based on the electrical requirements of the Envirogen treatment system.

For the demonstration at Port Hueneme, the Envirogen control panel system that utilized 110 volt power was mounted on a portable, unitrust assembly that was anchored on an exterior wall of the U.S. EPA shelter building. The demonstration system power was supplied by NETTS. The total annual electrical energy consumption provided by Envirogen was based on their project at a gas station in New Jersey with the total annual electricity costs are therefore estimated to be about $ 6,276.

Water and natural gas usage are highly site specific but assumed to be minimal for this analysis. As a result, no costs for these utilities are presented.

8.3.8 Effluent Treatment and Disposal Costs

Depending on the degree to which treatment goals for a site were met, additional effluent treatment may be required, and thus additional treatment or disposal costs may be incurred. Because of the uncertainty associated with additional treatment or disposal costs, this analysis does not include effluent treatment or disposal costs.

The Envirogen system requires air monitoring because of the application of propane as a treatment substrate. As a result, additional air emission control may be required, and thus additional treatment or disposal costs may be incurred. Because of the uncertainty associated with additional treatment costs, this analysis does not include effluent treatment costs.

However, it is assumed that effluent monitoring (ground water leaving the treatment zone) and the air emission at the vapor monitoring and the ground water points are routinely conducted by the primary operator.

8.3.9 Residual Waste Shipping and Handling Costs

The only residuals produced during a successful propane biostimulation and bioaugmentation system operation are fiber drums containing used PPE and waste sampling and field analytical supplies, all of which are typically associated with a ground-water project. This waste is assumed to be hazardous and requires disposal at a permitted facility. This analysis assumes that about six drums of waste are disposed of annually. The cost of handling and transporting the drums and disposing of them at a hazardous waste disposal facility is about $1,000 per drum. The total drum disposal costs are, therefore, about $6,000 per year.

8.3.10 Analytical Services Costs

Required sampling frequencies are highly site specific and are based on treatment goals and contaminant concentrations. Analytical costs associated with a ground-water treatment project include the costs of laboratory analyses, data reduction, and QA/QC. This analysis assumes that one sample of untreated (upgradient) water, one sample of treated water (downgradient), and associated QC samples (trip blanks, field duplicates, and matrix spike/matrix spike duplicates) will be analyzed for VOCs every month. Monthly analytical costs are estimated at $1,500. The total annual analytical costs are, therefore, estimated to be $18,000.

8.3.11   Equipment Maintenance Costs

Typically, annual equipment maintenance costs are estimated to about 3% of the capital equipment costs.

8.3.12   Site Demobilization Costs

Site demobilization includes treatment system shutdown, disassembly, and decontamination; site cleanup and restoration; utility disconnection; and transportation of the equipment off site. A two-person crew will work about five 8-hour days to disassemble and load the system. This analysis assumes that the equipment will be transported 1,000 miles either for storage or to the next job site. Generally, it is estimated that the total cost of demobilization is about $15,000. This total includes all labor, material, and transportation costs.

According to Table 8-1, the vendor stated that demobilization costs were minimal due to the proximity of the demonstration site to Envirogen's office. Elements of demobilization could include the following:

1.   Labor associated with equipment decommissioning and removal;
2.   Demobilization of staff;
3.   Subcontractor costs associated with abandonment of demonstration wells;
4.   Removal of above-grade distributions lines and equipment; and
5.   Site restoration.

Equipment decommissioning and removal and demobilization of staff at the NBVC demonstration site was accomplished in one-half day due to assistance from NBVC staff and would not be expected to exceed 3 days at the full scale.

8.4   CONCLUSIONS OF ECONOMIC ANALYSIS

This analysis presents cost estimates for treating contaminated ground water with the Envirogen treatment system at pilot scale for a typical gas station. Table 8-1 presents the total cost as provided by Envirogen for each cost category. Permitting and regulatory costs are not representative because they represent less than the normal costs. In addition, Effluent Treatment and Disposal Costs are not included in Table 8-1, since there are no cost estimates for this category. With Envirogen's assumptions (Section 8.2.3), the total cost to treat 81,000 gallons of contaminated ground water was estimated to be $2.35 per gallon.

In parallel to a cost estimate for a typical gas station, a cost estimate for the field demonstration at hazardous waste sites was presented. According to this analysis, one time costs (fixed costs) include site preparation, permitting and regulatory, mobilization and startup, equipment, and site demobilization. Annual costs include labor, supplies, utilities, effluent treatment and disposal, residual waste shipping and handling, analytical services, and equipment maintenance. This analysis of the technology shows that operating costs are strongly affected by the site specific environment, size and configuration of the vendor system, distance from the Envirogen office location with most of the annual costs per gallon being proportionally higher than estimated during this cost analysis, as presented in Table 8-1.

SECTION 9
TECHNOLOGY APPLICATIONS ANALYSIS

This section of the report addresses the general applicability of the Envirogen technology for treating contaminated ground water at hazardous waste and petroleum release sites. The analysis is based primarily on the demonstration results at the NBVC, and conclusions are based exclusively on these data since only limited information is available on full-scale applications of the technology. This demonstration was conducted over a ten-months period during June 2001 to March 2002. Vendor's claims regarding the effectiveness and applicability of the Envirogen technology are included in Appendix A.

This section also discusses the following topics regarding the applicability of the Envirogen technology: technology performance versus Applicable or Relevant and Appropriate Requirements (ARAR), technology operability, key features of the treatment technology, applicable wastes, availability and transportability of equipment, material handling requirements, range of suitable site characteristics, limitations of the technology, and potential regulatory requirements.

9.1   TECHNOLOGY PERFORMANCE VERSUS ARARS

The technology's ability to comply with existing federal, state, or local ARARs (for example, MCLs) should be determined on a site-specific basis. The discussion below focuses on the demonstration at the NBVC for MTBE-contaminated ground water.

For the MTBE technologies demonstration program at the NBVC, ARARs were identified and established by consensus among the stakeholders for the technology demonstration. ARARs included EPA and California Primary and Secondary MCLs for drinking water and California Public Health Goals for drinking water. For this demonstration, the contaminants initially present in the ground water which were of primary concern included MTBE and TBA (Table 2-3). Although TBA, a partially oxidized organic compound resulting from MTBE degradation was of concern, for the demonstration of this technology, MTBE was the only ground-water parameter that was identified as applicable.

In the demonstration at the NBVC, the Envirogen technology did not meet the treatment goals based on MCLs for the primary contaminants of concern. MTBE remained in the downgradient monitoring wells and was higher in concentration than potentially applicable ARARs.

108

In summary, according to the vendor, the Envirogen technology has been shown to be capable of reducing 85 percent of MTBE contaminants in ground water to below 70 µg/L in a pilot-scale study conducted at a gas station in New Jersey. For hydrocarbons, including BTEX and MTBE, compliance with MCLs may be problematic if BTEX compounds are the main contaminants of concern. Additionally, the presence of TBA, partially oxidized organic compound, may be of concern to ARAR compliance at specific sites, depending on the application and the planned disposal or reuse of the downgradient water from the Envirogen system. The following were identified as additional potential technology performance issues with respect to ARARs:

- The technology's ability to meet any future chemical-specific ARARs for by-products may be considered because of the potential for formation of TBA, partially oxidized organic during treatment.

- States may require SVE for system operation.

9.2 TECHNOLOGY OPERABILITY

The operation of the demonstration system involves the use of compressed gas cylinders to provide the source of oxygen and propane and simple timer-actuated solenoid valves to control flow. Therefore, the principal factor affecting Envirogen system performance is the delivery of the gases into solution. Tasks associated with the operation and maintenance of the system included routine flow and pressure measurements at the injection point, monitoring oxygen and propane use, and changing spent gas cylinders.

Propane biostimulation technology uses commercially available, off-the-shelf components to establish bioreactive treatment zones. Equipment used in the performance and monitoring of the demonstration is available through standard suppliers. The routine monitoring of the control system by the study participants indicated the use of a more sensitive control system would enhance optimum operating conditions.

Although Envirogen claims that their system can treat shallow aquifers, the presence of a deep water table could add to the cost and operating difficulties of the operation of the Envirogen technology. Also, as discussed earlier, the system would be less effective in aquifers with low hydraulic conductivities. The type of aquifers for which the Envirogen process is most effective include those composed of sand to cobbles and with hydraulic conductivities greater than 10^{-4} cm/sec. The irregular distribution of oxygen and propane caused by heterogeneities would result in zones where little or no treatment can occur.

Biochemical factors that must be present include microbes capable of degrading the contaminants of concern, the availability of nutrients, and a neutral pH. Other operating parameters that influence the performance of the Envirogen technology include the presence of excess propane.

9.3   KEY FEATURES OF THE TREATMENT TECHNOLOGY

Common methods for treating ground water contaminated with organic compounds include air stripping, steam stripping, carbon adsorption, biological treatment, and chemical oxidation. The Envirogen system is an in-situ technology that allows on-site treatment of contaminated ground water without excavation and with limited site preparation.

In situ treatment is advantageous, especially when volatile organic compounds are present since handling activities may be minimized. These technologies have the potential for the complete destruction of the contaminants rather than transferring them to another medium.

Envirogen operation involves injecting propane and oxygen into an MTBE-contaminated aquifer. The addition of these substrates promotes the growth of propane oxidizing bacteria (POB) and the production of the enzyme propane monooxygenous that catalyzes the complete destruction of MTBE and its primary daughter product, TBA. The injection of exogenous POB such as strain ENV 425 is used to seed the aquifer to insure activity or speed initiation of the treatment process.

The Envirogen system does not generate residue, sludge, or spent media that require further processing, handling, or disposal. As the target organic compound, MTBE is either mineralized or broken down into low molecular weight compounds. When complete destruction occurs, produced intermediate species are ultimately oxidized to CO_2 and water.

9.4   APPLICABLE WASTES

Based on Envirogen's claim, as well as results from a pilot scale demonstration at a gas station in New Jersey and other laboratory studies, the Envirogen technology may have applicability to treat MTBE in liquids, including ground water and wastewater. Where stringent effluent requirements apply, such as the demonstration at NBVC site, the technology does not appear to be particularly applicable to the treatment of contaminated ground waters containing MTBE. However, the technology can achieve substantial reductions in the concentrations of other petroleum hydrocarbons.

9.5   AVAILABILITY AND TRANSPORTABILITY OF EQUIPMENT

The vendor provides the complete Envirogen treatment system configured for site-specific conditions. All Envirogen treatment equipment is leased to the client. As a result, all depreciation and salvage value is incurred by Envirogen, which is reflected in the price for leasing the equipment. At the end of a treatment project, Envirogen decontaminates and demobilizes its treatment equipment. Envirogen assumes that this equipment will operate for the duration of the ground-water remediation project and will still function after the remediation is complete as a result of routine maintenance and modifications.

9.6   MATERIALS HANDLING REQUIREMENTS

Other than the soil cuttings generated during installation of the demonstration injection points, monitoring wells, and vapor monitoring points, and ground water derived from sampling during the demonstration, the Envirogen system does not generate treatment residuals that require further processing, handling, or disposal. Depending on various states, requirements, the Envirogen unit may require air emissions specific controls.

If MCLs are achieved, treated water then may be disposed of either on or off site, depending on site-specific requirements and limitations. Examples of on-site disposal options for treated water include ground-water recharge or temporary on-site storage for sanitary use. Examples of off-site disposal options include discharge into surface water bodies, storm sewers, and sanitary sewers. Bioassay tests may be required in addition to routine chemical and physical analyses before the treated water is disposed.

9.7   RANGE OF SUITABLE SITE CHARACTERISTICS

In addition to the quality of ground water entering the system and downgradient discharge requirements, site characteristics and support requirements are important when considering the Envirogen technology. Site-specific factors can impact the application of the Envirogen technology, and these factors should be considered before selecting the technology for remediation at a specific site. Site-specific factors addressed in this section include site support requirements and utility requirements.

111

According to Envirogen, both transportable and permanently installed Envirogen systems are available (see Section 10 Technology Status, and Appendix A, Vendor's Claims for the Technology). The support requirements for these systems are likely to vary. This section presents support requirements based on the information collected for the permanently installed system used during the demonstration.

9.7.1 Site Support Requirements

The main site requirement is the availability of electricity. For the unit used during the demonstration, a 3-phase, 206V power was utilized. The system controls operated using conditioned power reduced to 24V AC power to the individual timers and solenoid valves. These voltages are standard unit grid voltages available in the United States. Other utilities required for the use of Envirogen include water for cleaning; only small amounts of potable water are required. Access to the site must be provided over roads suitable for travel by heavy equipment. Personnel must also be able to reach the site without difficulty. An additional area is required for an office or laboratory building and for the storage of the equipment. A fence surrounds the Envirogen site to provide additional security. The fence should be posted with signs for "explosion hazard," and no smoking should be permitted anywhere on site. During the demonstration, an area of about 61 feet by 172 feet was used for the Envirogen plots, an EPA shed area, and miscellaneous equipment.

If the portable unit is used, the site must be accessible for a tractor-trailer truck with an 8-foot by 28-foot trailer weighing about 10 tons. An area 8 feet by 28 feet must be available for the trailer that houses the Envirogen system, and additional space must be available to allow personnel to move freely around the outside of the trailer. The area containing the Envirogen trailer should be paved or covered with compacted soil or gravel to prevent the trailer from sinking into soft ground.

9.8   LIMITATIONS OF THE TECHNOLOGY

Because the biostimulation technology is an extension of conventional air sparging and biosparging techniques, its application is generally limited by the same hydrologic factors that prevent conventional sparging. Sites that are characterized by low permeability formations, such as silts or clays, or heterogeneous soil conditions are the primary obstacles to successful treatment.

Another issue of concern for this technology is the risk of explosion caused by propane addition. Air sparging technologies have long been used for remediating gasoline contamination, thereby generating potentially explosive gaseous vapors of gasoline components and oxygen. In addition to water table mounding caused by the injection of treatment gases, system operation was limited in placing the gases into solution with the exception of those parts of the aquifer in proximity to the injection locations. During the time the propane injection system was operating, its odor was very pronounced at the surface.

Another limiting factor has been identified based on the principles of monoxygenous response to the mixture of contaminants. Based on research studies performed by Envirogen and demonstration results, the system is the most efficient if the MTBE concentrations significantly exceed the BTEX constituents in contaminated ground water being treated by the Envirogen system. If treatment goals are not met while the system operates, such a case would require operating additional Envirogen units in series, obtaining a larger Envirogen unit, or adding pretreatment or post-treatment, any of which would increase costs.

9.9   POTENTIAL REGULATORY REQUIREMENTS

This section discusses regulatory requirements pertinent to use of the Envirogen technology at Superfund and RCRA corrective action sites. The regulations applicable to implementation of this technology depend on site-specific remediation logistics and the type of contaminated liquid being treated; therefore, this section presents a general overview of the types of federal regulations that may apply under various conditions. State requirements should also be considered because they vary from state to state and, therefore, are not presented in detail in this section.

Depending on the characteristics of the ground water to be treated, pretreatment or post-treatment may be required for the successful operation of the Envirogen system. Each pretreatment or post-treatment process might involve additional regulatory requirements that would need to be determined in advance. No direct air emissions are generated by the Envirogen treatment process. Therefore, only regulations addressing contaminated ground-water treatment and discharge, potential fugitive air emissions, and additional considerations are discussed below.

Comprehensive Environmental Response, Compensation, and Liability Act (CERCLA)

The CERCLA of 1980, as amended by the Superfund Amendments and Reauthorization Act (SARA) of 1986, provides for federal funding to respond to releases or potential releases of any hazardous substance into the environment, as well as to releases of pollutants or contaminants that may present an imminent or significant danger to public health and welfare or to the environment.

As part of the requirement of CERCLA (EPA ,1988; 1989), the EPA has prepared the National Oil and Hazardous Substances Pollution Contingency Plan (NCP) for hazardous substance response. The NCP is codified in Title 40, Code of Federal Regulations (CFR), part 300, and delineates the methods and criteria used to determine the appropriate extent of removal and cleanup for hazardous waste contamination.

SARA states a strong statuary preference for innovative technologies that provide long-term protection and directs EPA to do the following:

- Use remedial alternatives that permanently and significantly reduce the volume, toxicity, or mobility of hazardous substances, pollutants, or contaminants;

- Select remedial options that protect human health and the environment, are cost-effective, and involve permanent solutions and alternative treatment or resource recovery technologies to the maximum extent possible; and

- Avoid off-site transport and disposal hazardous substances or contaminated materials when practicable treatment technologies exist [Section 121 (b)].

Although during this demonstration, the above stated criteria were not met, a successful in-situ technology would meet each of these requirements. In general, two types of responses are possible under CERCLA: removal and remedial action. Superfund remedial actions are conducted in response to an immediate threat caused by a release of hazardous substances. Removal action decisions are documented in an action memorandum. Many removals involve small quantities of waste or immediate threats requiring quick action to alleviate the hazard. Remedial actions are governed by the SARA amendments to CERCLA. As stated above, these amendments promote remedies that permanently reduce the volume, toxicity, and mobility of hazardous substances, pollutants, or contaminants.

On-site removal and remedial actions must comply with federal and often more stringent state ARARs. ARARs are determined on a site-by-site basis and may be waived under six conditions: (1) the action is an interim measure, and the ARAR will be met at completion; (2) compliance with the ARAR would pose a greater risk to health and the environment than noncompliance; (3) it is technically impracticable to meet the ARAR; (4) the standard of performance of an ARAR can be met by an equivalent method; (5) a state ARAR has not been consistently applied elsewhere; and (6) ARAR compliance would not provide a balance between the protection achieved at a particular site and demands on the Superfund for other sites.

These waiver options apply only to Superfund actions taken on-site, and justifications for the waiver must be clearly demonstrated.

Resource Conservation and Recovery Act

RCRA, an amendment to the Solid Waste Disposal Act (SWDA), is the primary federal legislation governing hazardous waste activities and was passed in 1976 to address the problem of how to safely dispose of municipal and industrial waste. Subtitle C of RCRA contains requirements for generation, transport, treatment, storage, and disposal of hazardous waste, most of which are also applicable to CERCLA activities. The Hazardous and Solid Waste Amendments of 1984 greatly expanded the scope and requirements of RCRA.

EPA and RCRA-authorized states (listed in 40 Code of Federal Regulations [CFR] Part 272) implement and enforce RCRA and state regulations. Some of the RCRA requirements under 40 CFR Part 264 Subpart F (promulgated) and Subpart S (partially promulgated) generally apply at Comprehensive Emergency Response, Compensation, and Liability Act (CERCLA) sites that contain RCRA hazardous waste because remedial actions generally involve treatment, storage, or disposal of hazardous waste. Subparts F and S include requirements for initiating and conducting RCRA corrective action, remediating ground water, and ensuring that corrective actions comply with other environmental regulations. Subpart S also details conditions under which particular RCRA requirements may be waived for temporary treatment units operating at corrective action sites and provides information regarding requirements for modifying permits to adequately describe the subject treatment unit.

According to Envirogen, the propane biostimulation technology can treat ground-water contaminated with petroleum hydrocarbons and MTBE. Contaminated ground water treated by the system may be classified as a RCRA hazardous waste or may be sufficiently similar to a RCRA hazardous waste that RCRA regulations will be applicable requirements.

Clean Water Act

The objective of the Clean Water Act (CWA) is to restore and maintain the chemical, physical, and biological quality of navigable surface waters by establishing federal, state, and local discharge standards. If treated ground water is discharged to surface water bodies or publicly owned treatment works (POTW), CWA regulations apply. On-site discharges to surface water bodies must meet substantive National

115

Pollutant Discharge Elimination System (NPDES) requirements but do not require an NPDES permit. A direct discharge of (CERCLA) wastewater would qualify as "onsite" if the receiving water body is in the area of contamination or in proximity to the site, and if the discharge is necessary to implement the response action. Off-site discharges to a surface water body require an NPDES permit and must meet NPD ES permit limits. Discharge to a POTW is considered to be an off-site activity, even if an on-site sewer is used. Therefore, compliance with substantive and administrative requirements of the National Pretreatment Program is required in such a case. General pretreatment regulations are included in 40 (CFR) Part 403.

Any applicable local or state requirements, such as local or state pretreatment requirements or water quality standards (WQS), must also be identified and satisfied. State WQSs are designed to protect existing and attainable surface water uses (for example, recreation and public water supply). WQSs include surface water use classifications and numerical or narrative standards (including effluent toxicity standards, chemical-specific requirements, and bioassay requirements to demonstrate no observable effect level [NOEL] from a discharge) (EPA 1988). These standards should be reviewed on a state- and location-specific basis before discharges are made to surface water bodies.

Safe Drinking Water Act

The Safe Drinking Water Act (SDWA) of 1974, as most recently amended by the Safe Drinking Water Amendments of 1986, required EPA to establish regulations to protect human health from contaminants in drinking water. EPA has developed the following programs to achieve this objective: (1) a drinking water standards program, (2) an underground injection control program, and (3) sole-source aquifer and well-head protection programs.

SDWA primary (or health-based) and secondary (or aesthetic) MCLs generally apply as cleanup standards for water that is, or may be, used as drinking water. In some cases, such as when multiple contaminants are present, more stringent maximum contaminant level goals (MCLG) may be appropriate. In other cases, alternate concentration limits (ACL) based on site-specific conditions may be applied. CERCLA and RCRA standards and guidance should be used in establishing ACLs (EPA 1987). During the demonstrations, Envirogen treatment system performance was tested for compliance with SDWA MCLs for MTBE as a critical VOC.

Water discharge through injection wells is regulated by the underground injection control program. Injection wells are categorized as Classes I through V, depending on their construction and use. Reinjection of treated water involves Class IV (reinjection) or Class V (recharge) wells and should meet SDWA requirements for well construction, operation, and closure. If the ground water treated is a RCRA hazardous waste, the treated ground water must meet RCRA Land Disposal Restriction (LDR) treatment standards (40 CFR Part 268) before reinjection.

The sole-source aquifer and well-head protection programs are designed to protect specific drinking water supply sources. If such a source is to be remediated using the Envirogen system, appropriate program officials should be notified, and any potential regulatory requirements should be identified. State ground-water antidegradation requirements and (WQSs) may also apply.

Clean Air Act

The Clean Air Act (CAA), as amended in 1990, regulates stationary and mobile sources of air emissions. CAA regulations are generally implemented through combined federal, state, and local programs. The CAA includes chemical-specific standards for major stationary sources that would not be applicable but could be relevant and appropriate for Envirogen system use. For example, because of the nature of the Envirogen process, which is a biosparging, the potential for stripping of VOCs and off-gassing of propane and oxygen may require SVE operation. Therefore, the Envirogen system may need to be controlled to ensure that air quality is not impacted. The National Emission Standards for Hazardous Air Pollutants (NESHAP) could also be relevant and appropriate if regulated hazardous air pollutants are emitted and if the treatment process is considered sufficiently similar to one regulated under these standards. In addition, New Source Performance Standards (NSPS) could be relevant and appropriate if the pollutant emitted and the Envirogen system are sufficiently similar to a pollutant and source category regulated by an NSPS. Finally, state and local air programs have been delegated significant air quality regulatory responsibilities, and some have developed programs to regulate toxic air pollutants (EPA 1989). Therefore, state air programs should be consulted regarding Envirogen treatment technology installation and use.

Toxic Substances Control Act

Testing, premanufacture notification, and record-keeping requirements for toxic substances are regulated under the Toxic Substances Control Act (TSCA). TSCA also includes storage requirements for polychlorinated biphenyls (PCB) (see 40 CFR §761.65). The Envirogen system may be used to treat ground water contaminated with PCBs, and TSCA requirements would apply to pretreatment storage of PCB-contaminated liquid. TCA was not an ARAR at the NBVC demonstration.

Occupational Safety and Health Act

OSHA regulations in 29 CFR Parts 1900 through 1926 are designed to protect worker health and safety. Both Superfund and RCRA corrective actions must meet OSHA requirements, particularly § 1910.120 which describes safety and health regulations for construction sites. On-site construction activities at Superfund or RCRA corrective action sites must be performed in accordance with 1926 of OSHA, which describes safety and health regulations for construction sites. For example, electric utility hookups for the Envirogen system must comply with Part 1926, Subpart K, Electrical.

In addition to meeting the OSHA requirements for the Envirogen treatment gases (for example, Part 1926, Subpart D, Occupational Health and Environmental Controls, and Subpart H, Materials Handling, Storage, and Disposal), all technicians operating the Envirogen system and all workers performing on-site work must have completed the OSHA training course and must be familiar with all OSHA requirements relevant to hazardous waste sites, in particular with those pertaining to the vendor's treatment gases, oxygen and propane, material safety data information as stated in 29 CFR 1910, Subpart Z. State OSHA requirements, which may be more stringent than federal standards, must also be met. In addition, health and safety plans for site remediations should address chemicals of concern and include monitoring practices to ensure that worker health and safety are maintained.

State and Community Acceptance

Because few applications of the Envirogen technology have been attempted beyond the bench or pilot scale, limited information is available to assess state and community acceptance of the technology. During the demonstrations at the NBVC, more than 100 people from regulatory agencies, nearby universities, and the local community attended Visitors' Day to observe demonstration activities and ask questions pertaining to the technology. The visitors expressed no concerns regarding operation of the Envirogen system.

SECTION 10
TECHNOLOGY STATUS

Envirogen claims that their technology can be used for the remediation of contaminated ground water, especially when contaminated with MTBE and TBA. However, at the NBVC demonstration site, the Envirogen technology failed to remove MTBE to the compliance level of 5 µg/L. There are currently no commercially operating systems in the United States.

The equipment and materials necessary to install the Envirogen technology are readily available. Prior to installation, the subsurface hydrogeology, waste distribution, waste characteristics, and ground-water chemistry must be characterized. Envirogen uses a three-phase approach in implementing its propane biostimulation technology. During phase 1, a bench-scale treatability study is performed using aquifer materials and a small quantity of ground water to construct microcosms. The purpose of this phase is to determine the abundance of indigenous bacteria at a site and their potential for the removal of MTBE. The results from microcosm studies are also used to determine if bioaugmentation is required. During phase 2, a pilot-scale treatability study is conducted on site using Envirogen's trailer mounted system. The results of this study will be used to (1) determine the effectiveness of the technology under site conditions, (2) design a full-scale system to meet treatment goals, and (3) provide duration and cost estimates for full-scale system operation. During phase 3, a full-scale operation of the technology will be implemented.

10.1   PREVIOUS EXPERIENCE

In addition to this demonstration, Envirogen's bioaugmentation and biostimulation technology has been applied for the remediation of MTBE contaminated ground water at an operating gasoline service station in New Jersey. During that project, which was conducted following an air sparging project, Envirogen utilized pre-existing sparging and SVE wells, with slight modification, for the injection of oxygen and propane as well as capturing the excess of propane. Envirogen claims that, as a result of the application of their technology at this service station, a significant amount of MTBE was reduced. Envirogen further claims that, although the TBA concentrations in the ground water increased during MTBE degradation, it was orders of magnitude lower than the MTBE concentrations.

119

10.2 SCALING CAPABILITITES

The specifics of the components of the system utilized during this demonstration are provided in Section 1.3 and the cost per unit estimate is provided in Section 8. Additionally, the vendor claims that a variety of the systems from a small portable to the large permanent systems are available to accommodate sites with different volumes of contaminated ground water. Since the systems are modular in nature, once the treatment design is completed, installation of the equipment can take from one week to one month depending on regulatory requirements, the number of injection wells, and the complexity of the treatment system.

REFERENCES

Aeschbach-Hertig, W., Schlosser, P., Stute, M, Simpson, H. J., Ludin, A., and Clark, J. F. 1998. A ^{3}H/^{3}He Study of Ground Water Flow in a Fractured Bedrock Aquifer. Ground Water 36(4): 661-670.

American Water Works Association (AWWA). 1998. Standard Method for the Examination of Water and Wastewater, 20th Edition. American Public Health Association, American Water Works Association and Water Environmental Federation.

Benner, M. L., Stanford, S. M., Lee, L. S., and Mohtar, R. H. 2000. Field and Numerical Analysis of In-Situ Air Sparging: A Case Study. J. Hazard. Mater. 72(2-3): 217-236.

Bernauer, U., Amberg, A., Scheutzow, D., and Dekant, W. 1998. Biotransformation of ^{12}C- and 2-^{13}C-Labeled Methyl *tert*-Butyl Ether, Ethyl *tert*-Butyl Ether, and *tert*-Butyl Alcohol in Rats: Identification of Metabolites in Urine by ^{13}C Nuclear Magnetic Resonance and Gas Chromatography/Mass Spectrometry. Chem. Res. Toxicol. 11: 651-658.

Bianchi, A. and Varney, M. S. 1989. Analysis of Methyl *tert*-Butyl Ether and 1, 2-Dihaloethanes in Estuarine Water and Sediments Using Purge-and-Trap/Gas-Chromatography. J. High Resolut. Chromatogr. 12: 184-186.

Bonin, M. A., Ashley, D. L., Cardinali, F. L., McCraw, J. M., Wooten, J. V. 1995. Measurement of Methyl *tert*-Butyl Ether and *tert*-Butyl Alcohol in Human Blood by Purge-and-Trap Gas Chromatography- Mass Spectrometry Using an Isotope-Dilution Method. J. Anal. Toxicol. 19: 187-191.

Borden, R. C., Daniel, R. A., LeBrun, L. E., and Davis, C. W. 1997. Intrinsic Biodegradation of MTBE and BTEX in a Gasoline-Contaminated Aquifer. Water Resour. Res. 33(5): 1105-1115.

Bowman, R. S. and Gibbens, J. F. 1992. Difluorobenzoates as Nonreactive Tracers in Soil and Ground Water. Ground Water 30(1): 8-14.

Bullivant, D. P., and O'Sullivan, M. J. 1989. Matching a Field Tracer Test with some Simple Models. Water Resour. Res. 25(8): 1879-1891.

Church, C. D., Isabelle, L. M., Pankow, J. F., Rose, D. L., and Tratnyek, P. G. 1997a. Method for Determination of Methyl *tert*-Butyl Ether and its Degradation Products in Water. Environ. Sci. Technol. 31(12): 3723-3726.

Church, C. D., Isabelle, L. M., Pankow, J. F., Tratnyek, P. G., and Rose, D. L. 1997b. Assessing the In Situ Degradation of Methyl *tert*-Butyl Ether (MTBE) by Product Identification at the Sub-PPB Level Using Direct Aqueous Injection GC/MS. Division of Environmental Chemistry Preprint of Extended Abstracts 37(1): 411-413.

Church, C. D., Pankow, J. F., and Tratnyek, P. G. 1999a. Hydrolysis of *tert*-Butyl Formate: Kinetics, Products, and Implications for the Environmental Impact of Methyl *tert*-Butyl Ether. Environ. Toxicol. Chem. 18(12): 2789-2796.

Church, C. D., Tratnyek, P. G., Pankow, J. F., Landmeyer, J. E., Baehr, A. L., Thomas, M. A., and Schirmer, M. 1999b. Effects of Environmental Conditions on MTBE Degradation in Model Column Aquifers. U. S. Geological Survey, Water Resources Investigations Report 99-4018C, 3: 93-101.

Church, C. D., Tratnyek, P. G., and Scow, K. M. 2000. Pathways for the Degradation of MTBE and Other Fuel Oxygenates by Isolate PM1. Am. Chem. Soc. 40(1): 261-263.

Clayton, W. S., Brown, R. A., and Bass, D. H. 1995. Air Sparging and Bioremediation: The Case for In Situ Mixing. In Situ Aeration: Air sparging, bioventing, and related remediation processes 3(2): 75-85.

Code of Fedral Regulations (CFR). 1002. Title 40. Part 136.

Connell, L. D. 1994. The Importance of Pulse Duration in Pulse Test Analysis. Water Resour. Res. 30(8): 2403-2411.

Davis, G. B., Patterson, B. M., Thierrin, J., Benker, E. 2000. Deuterated Tracers for Assessing Natural Attenuation in Contaminated Groundwater. In Proceedings of Tracers and Modeling in Hydrogeology of the TraM'2000 Conference in Liege, Belgium. IAHS no. 262: 241-247. Englewood Cliffs, New Jersey.

EPA. 1987. Alternate Concentration limit (ACL) Guidance. Part 1: ACL Policy and Information Requirements. EPA/530/SW-87/017.

EPA. 1988. CERCLA Compliance with Other Environmental Laws: Interim Final. OSWER. EPA/540/G-89/006.

EPA. 1989. CERCLA Compliance with Other Laws Manual: Part II. Clean Air Act and Other Environmental Statutes and State Requirements. OSWER. EPA/540/G-89/006.

EPA. 1995. Methods for the Determination of Organic Compounds in Drinking Water, EPA 600/4-88/039. With Supplements II (1992) and III (1995).

EPA. 1996. Test Methods for Evaluating Solid Waste, Physical/Chemical Methods, Laboratory Manual, Volumes 1A through 1C, and Field Manual, Volume 2. SW-846, Third Edition and Update III, EPA document control no. 955-001-0000-1 Office of Solid Waste. September.

Evans, G. 1990. Estimating Innovative Treatment Technology Costs for the SITE Program. J. Air Waste Management Assoc. 40(7): 69-83, July.

Everett, L. G., Cullen, S. J., Rice, D. W., McNab Jr., W. W. Dooher, B. P., Kavanaugh, M. C., Johnson, P. C., Kastenberg, W. E., Small, M. C. 1998. Risk-Based Assessment of Appropriate Fuel Hydrocarbon Cleanup Strategies for the Naval Exchange Gasoline Station Naval Construction Battalion Center Port Hueneme, California. UCRL-AR-130891.

Garnier, P. M., Auria, R., Augur, C., and Revah, S. 1999. Cometabolic Biodegradation of Methyl t-Butyl Ether by *Pseudomonas Aeruginosa* Grown on Pentane. Appl. Microbiol. Biotechnol. 51: 489-503.

Gullick, R. W., and LeChevallier, M. W. 2002 Occurrence of MTBE in Drinking Water Sources. J. Am. Water Works Assoc. 92(1): 100-113.

Gupta, S. K., Lau, L. S., and Moravcik, P. S. 1994. Ground-Water Tracing with Injected Helium. Ground Water 32(1): 96-102.

Hanson, J. R., Ackerman, C. E., and Scow, K. M. 1999. Biodegradation of Methyl *tert*-Butyl Ether by a Bacterial Pure Culture. Appl. Environ. Microbiol. 65(11): 4788-4792.

Hardison, L. K., Curry, S. S., Ciuffetti, L.M., and Hyman, M. R. 1997. Metabolism of Diethyl Ether and Cometabolism of Methyl *tert*-Butyl Ether by a Filamentous Fungus, a Graphium sp. Appl. Environ. Microbiol. 63(8): 3058-3067.

Hyman, M., Kwon, P., Williamson, K., and O'Reilly, K. 1998. Cometabolism of MTBE by Alkane-Utilizing Microorganisms. In G. Wickramanayake, B. Hinchee, E. Robert (Eds.), Natural Attenuation; Chlorinated and Recalcitrant Compounds, 3: 321-325. Battelle Press, Columbus, OH.

Jensen, H. M. and Arvin, E. 1990. Solubility and Degradability of the Gasoline Additive MTBE, Methyl-*tert*.-butyl-ether, and Gasoline Compounds in Water. Contam. Soil: 445-448.

Ji, W., Dahmani, A., Ahlfeld, D. P., Lin, J. D., Hill III, E. 1993. Laboratory Study of Air Sparging: Air Flow Visualization. GWMR. 13(4): 115-126.

Johnson, P. C., Johnson, R. L., Neaville, C., Hansen, E. E., Stearns, S. M., and Dortch, I. J. 1997. An Assessment of Conventional In Situ Air Sparging Pilot Tests. Ground Water 35(5): 765-774.

Johnson, R. L. 1994. Enhancing Biodegradation with In Situ Air Sparging: A Conceptual Model. Air Sparging for Site Remediation 2(5): 14-22.

Johnson, R. L., Johnson, P. C., McWhorter, D. B., Hinchee, R. E., and Goodman, I. 1993. An Overview of In Situ Air Sparging. GWMR: 127-135.

Kanal, H., Inouye, V., Goo, R., Chow, R., Yazawa, L., and Maka, J. 1994. GC/MS Analysis of MTBE, ETBE, and TAME in Gasoline. Anal. Chem. 66(6): 924-927.

Kenoyer, G. J. 1988. Tracer Test Analysis of Anisotropy in Hydraulic Conductivity of Granular Aquifers. GWMR: 67-70.

Keppel, Geoffrey, *Design and Analysis: A Researcher's Handbook*, 1982, Prentice-Hall, Inc., New York, NY.

Kostecki, P. T. Calabrese, E. J., Bonazountas, M. 1997. Contaminated Soils Volume 2. Site Assessment Chemical Analysis and Environmental Fate Risk Assessment, Remediation Bioremediation, State Regulatory Federal and Military Considerations Federal and Military Considerations Manufactured Gas Plant Sites, Radioactivity, MTBE. Amherst Scientific Publishers. 661-679.

Meiri, D. 1989. A Tracer Test for Detecting Cross Contamination Along a Monitoring Well Column. GWMR: 78-81.

Melville, J. G., Molz, F. J., Guven, O., and Widdowson, M. A. 1991. Multilevel Slug Tests with Comparisons to Tracer Data. Ground Water 29(6): 897-907.

Mo, K., Lora, C. O., Wanken, A. E., Javanmardian, M., Yang, X., and Kulpa, C. F. 1999. Biodegradation of Methyl T-Butyl Ether by Pure Bacterial Cultures. Appl. Microbiol. Biotechnol. 47: 69-72.

Nouri, B., Fouillet, B., Toussaint, G., Chambon, R., and Chambon, P. 1996. Complementary of Purge-and-Trap and Head-space Capillary Gas Chromatographic Methods for Determination of Methyl-*tert*.-Butyl Ether in Water. J. Chromatogr. A, 726: 153-159.

Pankow, J. F., Johnson, R. L., and Cherry, J. A. 1993. Air Sparging in Gate Wells in Cutoff Walls and Trenches for Control of Plumes of Volatile Organic Compounds (VOCs). Ground Water 31(4): 654-663.

Pankow, J. F., Thomson, N. R., Johnson, R. L., Baehr, A. L., and Zogorski, J. S. 1997. The Urban Atmosphere as a Non-Point Source for the Transport of MTBE and Other Volatile Organic Compounds (VOCs) to Shallow Groundwater. Environ. Sci. Technol. 31(10): 2821-2828.

Parker, J. C., and van Genuchten, M. Th. 1984. Determining Transport Parameters from Laboratory and Field Tracer Experiments. Virginia Agricultural Experiment Station, Virginia Polytechnic Institute and State University Bulletin 84-3: 1-97.

Patrick, G. C., and Barker, J. F. 1985. A Natural-Gradient Tracer Study of Dissolved Benzene, Toluene and Xylenes in Ground Water. Second Canadian/American Conference on Hydrogeology. 141-147.

Poulson, S. R., Drever, J. I. and Colberg, P. J. S. 1997. Estimation of K_{OC} Values for Deuterated Benzene, Toluene, and Ethylbenzene, and Application to Ground Water Contamination Studies. Chemosphere 35: 2215-2224.

Poulson, S. R., Ohmoto, H. and Thomas, P. R. 1995. Stable Isotope Geochemistry of Waters and Gases (CO_2, CH_4) from the Overpressured Morganza and Moore-Sams Fields, Louisiana Gulf Coast. Appl. Geochem. 10: 407-417.

Reuter, J. E., Allen, B. C., Richards, R. C., Pankow, J. F., Goldman, C. R., Scholl, R. L., and Seyfried, J. S. 1998. Concentrations, Sources, and Fate of the Gasoline Oxygenate Methyl tert-Butyl Ether (MTBE) in a Multiple-Use Lake. Environ. Sci. Technol. 32(23): 3666-3672.

Rice, D. W., Grose, R. D., Michaelsen, J. C., Dooher, B. P., MacQueen, D. H., Cullen, S. J., Kastenberg, W. E., Everett, L. G., Marino, M. A. 1995. California Leaking Underground Fuel Tank (LUFT) Historical Case Analyses. California State Water Resources Control Board Underground Storage Tank Program and the Senate Bill 1764 Leaking Underground Fuel Tank Advisory Committee UCRL-AR-122207.

Ronen, D., Berkowitz, B., and Magaritz, M. 1993. Vertical Heterogeneity in Horizontal Components of Specific Discharge: Case Study Analysis. Ground Water 31(1): 33-40.

Salanitro, J. P., Johnson, P. C., Spinnler, G. E., Maner, P. M., Wisniewski, H. L., and Bruce, C. 2000. Field-Scale Demonstration of Enhanced MTBE Bioremediation Through Aquifer Bioaugmentation and Oxygenation. Environ. Sci. Technol. 34: 4152-4162.

Salanitro, J. P., Spinnler, G. E., Neaville, C. C., Maner, P. M., Stearns, S. M., Johnson, P. C., and Bruce, C. 1999. Demonstration of the Enhanced MTBE Bioremediation (EMB) In Situ Process. In-Situ Bioremediation of Petroleum Hydrocarbon and other Organic Compounds 5(3): 37-46.

Schirmer, M., Butler, B. J., Barker, J. F., Church, C. D., and Schirmer, K. 1999. Evaluation of Biodegradation and Dispersion as Natural Attenuation Processes of MTBE and Benzene at the Borden Field Site. Phys. Chem. Earth (B) 24(6): 557-560.

Smith, R. L., Harvey, R. W., and LeBlanc, D. R. 1991. Importance of Closely Spaced Vertical Sampling in Delineating Chemical and Microbiological Gradients in Groundwater Studies. J. Contamin. Hydrol. 7: 285-300.

Stevenson, D., Paling, W. A. J. and De Jesus, A. S. M. 1989. Radiotracer Dispersion Tests in a Fissured Aquifer. J. Hydrol. 110: 153-164.

Stute, M., Deak, J., Revesz, K., Bohlke, J. K., Deseo, E., Weppernig, R., and Schlosser, P. 1987. Tritium/3He Dating of River Infiltration: An Example from the Danube in Szigetkoz Area, Hungary. Ground Water 35(5): 905-911.

Suflita, J. M., and Mormile, M. R. 1993. Anaerobic Biodegradation of Known and Potential Gasoline Oxygenates in the Terrestrial Subsurface. Environ. Sci. Technol. 27(5): 976-978.

SYSTAT, 1990. The System for Statistics, Evanston, IL, SYSTAT, Inc.

Thierren, J., Davis, G. B., Barber, C., and Power, T. R. 1992. Use of Deuterated Organic Compounds as Groundwater Tracers for Determination of Natural Degradation Rates Within a Contaminated Zone. In H. Hotzl and A. Werner (Eds.), Tracer Hydrology, 85-91. A. A. Balkema, Rotterdam.

Thierrin, J., Davis, G. B., and Barber, C. 1995. A Ground-Water Tracer Test with Deuterated Compounds for Monitoring In Situ Biodegradation and Retardation of Aromatic Hydrocarbons. Ground Water 33(3): 469-475.

Thierrin, J., Davis, G. B., Barber, C., Patterson, B. M., Pribac, F., Power, T. R., and Lambert, M. 1993. Natural Degradation Rates of BTEX Compounds and Naphthalene in a Sulphate Reducing Groundwater Environment. J. Sci. Hydrol. 38(4): 309-322.

White, G. F., Russell, N. J., and Tidswell, E. C. 1996. Bacterial Scission of Ether Bonds. Microbiol. Rev. 60(1): 216-232.

Yen, C. K., and Novak, J. T. 1994. Anaerobic Biodegradation of Gasoline Oxygenates in Soils. Water Environ. Res. 66(5): 744-752.

APPENDIX A
VENDOR'S CLAIMS

This appendix was generated and written solely by Envirogen. The statements presented herein represent the vendor's point of view and summarize the claims made by the vendor, Envirogen (Lawrenceville, New Jersey), regarding their in-situ propane biostimulation technology. Publication herein does not represent the EPA's approval or endorsement of the statements made in this section; the EPA's point of view is discussed in the body of this report.

A.1   Introduction

MTBE has been used extensively as a gasoline additive in the United States to enhance combustion efficiency and reduced vehicle emissions, and its widespread use has ultimately led to its accidental release in the environment. Because it is present in high concentrations in reformulated gasoline and highly soluble in groundwater (K_{ow} 1.05), even small releases of gasoline can result in large MTBE plumes. The incidence of spills of MTBE-containing fuels from confirmed leaking underground storage tanks (USTs) in the United States has been estimated to be as high as 250,000. Sites contaminated with MTBE can vary in size from large terminals owned by multinational corporations to small family-owned service stations located near residential neighborhoods. Remedial technologies for treating MTBE, therefore, must be efficient, cost effective, and adaptable to a wide range of site conditions and limitations. Traditional remedial technologies such as activated carbon adsorption and air-sparging have proven to be largely ineffective or expensive for treating MTBE contamination, and it is clear that no single technology is suitable for every contaminated site. Recently, bioremediation has emerged as a suitable remedial alternative for some sites, and it can be applied by stimulating indigenous MTBE-degrading bacteria, or by adding exogenous bacteria, depending on conditions at the target site.

A.2   Biostimulation Technology Description

Biostimulation is a process by which the degradative activity of indigenous or added microorganisms is enhanced by adding specific nutrients or co-substrates that might otherwise be lacking or limiting. Often, indigenous microbes can be stimulated simply by adding a missing terminal electron acceptor like oxygen. Because some contaminants are not good growth substrates for indigenous bacteria, biostimulation sometimes can be facilitated by adding a co-metabolic growth substrates. Co-metabolism is a process by which the same enzyme that degrades a good growth substrate also fortuitously degrades the contaminant, often with little or no benefit to the degradative organisms. We demonstrated that propane oxidizing bacteria can co-metabolically mineralize MTBE to CO_2 and H_2O after growth on propane (Steffan et al., 1997). Because other hydrocarbon gases, such as methane and butane, have been used to stimulate co-metabolic biodegradation processes in situ, it is likely that a similar application of biostimulation, whereby propane and oxygen are injected to stimulate MTBE degradation by indigenous organisms or seed cultures, is feasible at some sites (US Patent # 5,814,514, Sept. 29, 1998).

There are several potential advantages to using a co-metabolic biostimulation approach for degrading MTBE in situ. Co-metabolism uncouples biodegradation of the contaminant from growth of the organisms. That is, the microbes can be supplied sufficient co-substrate (e.g., propane) to support growth, so they do not have to rely on the utilization of low levels of contaminants to maintain their survival. Also, the technology can be applied in a number of configurations depending on site characteristics and treatment needs. Possible application scenarios include: 1) re-engineered or modified multi-point AS/SVE systems that deliver propane and air throughout a contaminated site (suitable for use with existing AS/SVE systems or specially designed systems); 2) a series of air/propane delivery points arranged to form a permeable treatment wall to prevent off site migration of MTBE; 3) permeable treatment trenches fitted with air and propane injection systems; 4) in situ recirculating treatment cells that rely on pumping and reinjection to capture and treat a migrating contaminant plume; and 5) propane and oxygen injection through bubble-free gas injection devices to minimize off-gas release and contaminant stripping. Furthermore, propane is widely available, transportable even to remote sites, already present at many gasoline stations, and relatively inexpensive. Thus, propane biostimulation has the potential to be an attractive remediation option at a wide variety of MTBE-contaminated sites.

A.3   Demonstration results

During this project, we applied and evaluated propane biostimulation for MTBE remediation at the Port Hueneme, CA National Environmental Technologies Test Site. The primary purposes of this field demonstration included:

- Evaluating the effectiveness of propane biostimulation for MTBE remediation
- Optimizing sparging and SVE flow rates and injection/extraction cycles;
- Quantitatively assessing the impact of propane sparging on soil gas and ambient air quality;
- Delineating the zone of influence of the treatment;

- Assessing the potential for subsurface gas migration and fugitive emissions; and,
- Assess our ability to degrade MTBE to less than 5 µg/L with a single row of propane and oxygen injection points.

Microcosm testing with samples from the site revealed that the resident groundwater had low indigenous MTBE degrading microbial activity, even though MTBE degradation by native organisms has been observed during other demonstrations near our test plots. Consequently, we elected to seed our Test plot with a seed culture of propane oxidizing bacteria to initiate biodegradation

During the demonstration MTBE was degraded in both our Test (propane, oxygen, and bacteria added) and Control Plot (no propane added), but in neither case were the MTBE concentrations maintained at below the desired level of 5 µg/L. However, low levels of MTBE were achieved in many of the monitoring wells. For example, MTBE concentrations in the first row of deep Test Plot monitoring wells, GWT-2D, GWT-3D, and GWT-4D, went from 850, 1440, and 1440 µg/L at the beginning of the treatment (6/12/01) to 19, 46, and 440 µg/L at the end of treatment (3/12/02), respectively. Mean MTBE concentrations in the second row of monitoring wells went from 1967 µg/L (+/- 556 µg/L; n=3) to 148 µg/L (+/- 88 µg/L; n=3) during the same period. Likewise, MTBE concentrations of <5 µg/L were achieved in at least two of the shallow monitoring wells in the test plot. These low levels were achieved despite the addition of dMTBE as a tracer by the EPA which increased the total load of MTBE to the test plots. Variability in groundwater flow through the plots, and temporally during the course of the demonstration, appeared to affect distribution of co-substrates and oxygen in the test plot, and it made it difficult to accurately quantify the extent of MTBE degradation in the plots.

At the end of the field demonstration, experiments were performed to isolate MTBE degrading organisms from both the Test and Control Plot. Enrichment culturing with propane as a carbon source allowed growth of propane/MTBE degrading microorganisms from the Test Plot, but not from the Control Plot. Isolated propanotrophs from the Test Plot were phenotypically different (colony morphology and color) than the *Rhodococcus ruber* ENV425 culture added to the aquifer. Organisms able to grow on MTBE as a sole carbon source were isolated from both plots. These results suggest that the addition of propane to the Test Plot did allow growth of indigenous propane oxidizing microorganisms that were able to degrade MTBE. Similarly, addition of oxygen to both plots appeared to stimulate the growth of indigenous microbes capable of growth on MTBE.

Response to oxygen addition in the Control Plot was more rapid than anticipated based on microcosm studies performed by us, and based on prior demonstrations at the site. This high level of activity frustrated analysis of the effect of propane biostimulation on MTBE degradation at the site. Likewise, changes in the groundwater flow also made analysis of the data difficult. For example, because degradation rate calculations are dependent on groundwater flow, and because the hydraulic gradient was flat and the flow was low at the site, even small variations in flow could significantly affect degradation rate calculations. Groundwater elevation data even suggested that groundwater flow may have reversed its flow direction periodically during the study, especially in the Test Plot. Thus, unlike our prior demonstration where the positive effects of propane biostimulation were obvious (see below) the effects are less apparent in results of this study.

This demonstration also demonstrated that propane biosparging can be safely and economically applied at the field scale. Application of the technology resulted in no measurable fugitive emissions of propane, and in situ biodegradation and controlled propane addition maintained propane levels near or below its detection limit in groundwater. Propane costs for the 10-month demonstration were only about $50/month, indicating that application of this technology costs little more than a traditional air sparging system. Because of low propane emissions, the technology should not require secondary containment systems (e.g., soil vapor extraction) in most cases. Thus, it may be cost effective to incorporate propane biosparging equipment into MTBE remediation designs, even at sites where MTBE biodegradation by indigenous organisms is suspected. If indigenous bacteria prove to be inefficient or ineffective at remediating the site, propane can be injected to enhance activity at minimal additional cost.

Results of this demonstration also suggested that most of the active MTBE degradation that occurred in both plots occurred near the oxygen injection points. Thus, degradation activity may have been limited by the availability of oxygen in the subsurface. Oxygen was likely consumed by both geochemical oxygen sinks and biological activity. Because of the process monitoring and technology validation procedures of both Envirogen and the EPA, we elected not to increase gas flows into the site during this demonstration. To reach even lower MTBE levels, however, either additional rows of oxygen injection points or higher oxygen loading rates may be needed.

A.4   Case Study

Introduction. Propane biostimulation for MTBE remediation was applied at an operating Camden County, New Jersey service station site. A site investigation was initiated at the site after one of the site's underground gasoline tanks failed a tightness test in July 1988. The site has since undergone a range of remedial actions including soil excavation and air sparging. Six on-site groundwater monitoring wells (MW-5 to MW-10) and two offsite wells (MW-11 and MW-12) were installed to monitor BTEX and MTBE (Figure A-1). These wells are currently being monitored on a quarterly basis. Groundwater samples collected on February 9, 1999 showed site MTBE concentrations ranging from 170 µg/L (at upgradient monitoring well MW-8) to 270,000 µg/L (MW-6). Historical groundwater MTBE data from 1990 to 1999 indicate increasing concentrations at monitoring wells MW-6, MW-7, MW-9 and MW-11.

Because of the failure of air sparging and soil vapor extraction to sufficiently remove MTBE from the site groundwater, Envirogen was asked to perform propane biostimulation at the site. A biosparging and propane injection system was designed to allow flexible and safe implementation in the field. The system consisted of injection and SVE components, and utilize existing sparge wells (SP-1, SP-2 and SP-3) and SVE wells (VP-1, VP-2 and MW-10) at the site. The injection system consisted of two separate components; an air compressor and a propane supply system that was connected to the existing sparging distribution lines via a common manifold. An in-line filter was installed on the injection line to remove moisture and/or oil escaping the air compressor. The SVE system consisted of a vacuum blower that was connected to the existing SVE distribution lines and a carbon canister for treatment of the off-gas. Operation of the system was controlled using a common control panel with redundant control switches to ensure safe operations. An interlock devise was used to prevent propane injection unless the SVE system was operational.

Because the existing air sparging wells were not designed and constructed for pulsed operation, operation of the wells in a pulsed mode resulted in an accumulation of silt in the wells and reduced airflow. Consequently, the sparging system was operated with a continuous low airflow of 13 scfm. A 10-pound propane gas cylinder (e.g., similar in size to home barbecue propane tanks) was used as the propane supply. The discharge from the propane cylinder was controlled by a flow valve and pressure indicator mounted on the cylinder. A pressure control valve set at 40 psi was utilized to monitor and control the propane pressure in the line. An in-line propane lower explosive limit (LEL) detector was installed to continuously monitor the LEL level and ensure safe operation of the system. Dedicated flow meters were installed on each line to control the flow to each sparge well. Propane was added to the air stream for 10 min every three hours at a rate that ensured that the propane concentration did not exceed 0.2% propane in air (10% of the propane LEL). Approximately 0.5 lbs. of propane and 315 lbs. of oxygen were added to the site each day.

Results. Preliminary laboratory studies revealed that the site had low levels of indigenous microorganisms, presumably because of the low groundwater pH (pH ~3.5 to 5). Therefore, we elected to seed the aquifer with *R. ruber* ENV425. The system was initially operated for approximately one month without the addition MTBE degrading microorganisms. A total of 17 L of culture of strain ENV425 (~ 1 x 10^{11} cells/ml) was then added to the three sparge points. Bacterial injection was followed by several cycles of air sparging to help distribute the microbes into the treatment zone, and two days of continuous propane and air sparging to aid in establishing an active MTBE degrading microbial population. Because the low measured pH in ground water at the site, the ground water needed to be buffered to raise the pH to a range more favorable to MTBE biodegradation. A buffer solution of sodium bicarbonate was added to the sparge point periodically during the demonstration to achieve this goal. During each buffering event, a total of 120 gallons of a sodium bicarbonate solution was added to the sparge points followed immediately by air sparging to disperse the buffer into the formation.The system was operated for an additional ~5 months before a scheduled shutdown. MTBE and BTEX concentrations in the groundwater were measured using EPA method 8260.

Groundwater monitoring during the project was performed in monitoring wells MW-6, MW-7, MW-9, and MW-11 (Figure A-1). MW-6 is located just upgradient of the treatment zone, but it was slightly influenced by the treatment as indicated by increased dissolved oxygen in the groundwater during the treatment system operation. MW-7 also was upgradient of the treatment wells, but clearly within the zone of influence of the propane and oxygen injection system. MW-9 was immediately down gradient of the sparging points, and MW-11 was far down gradient of the treatment system.

MTBE concentrations in MW-6 were reduced by approximately 40% during the 5-month treatment period (Figure A-2A). Likewise, MTBE concentration is MW-7 were reduced by as much as 76% during biostimulation treatment (Figure A-2B). MTBE concentrations in MW-9 were reduced by as much as 98%, from 88 mg/L to 1.7 mg/L, during the treatment period (Figure A-2C). MTBE concentrations in MW 11 were relatively constant during the 5-month treatment period (data not shown), presumably because it was too far down gradient for treated water did not reach it during the demonstration period. First order rate constants for MW-6, MW-7 and MW-9 were calculated to be 0.0084, 0.0288, and 0.0027/day, respectively.

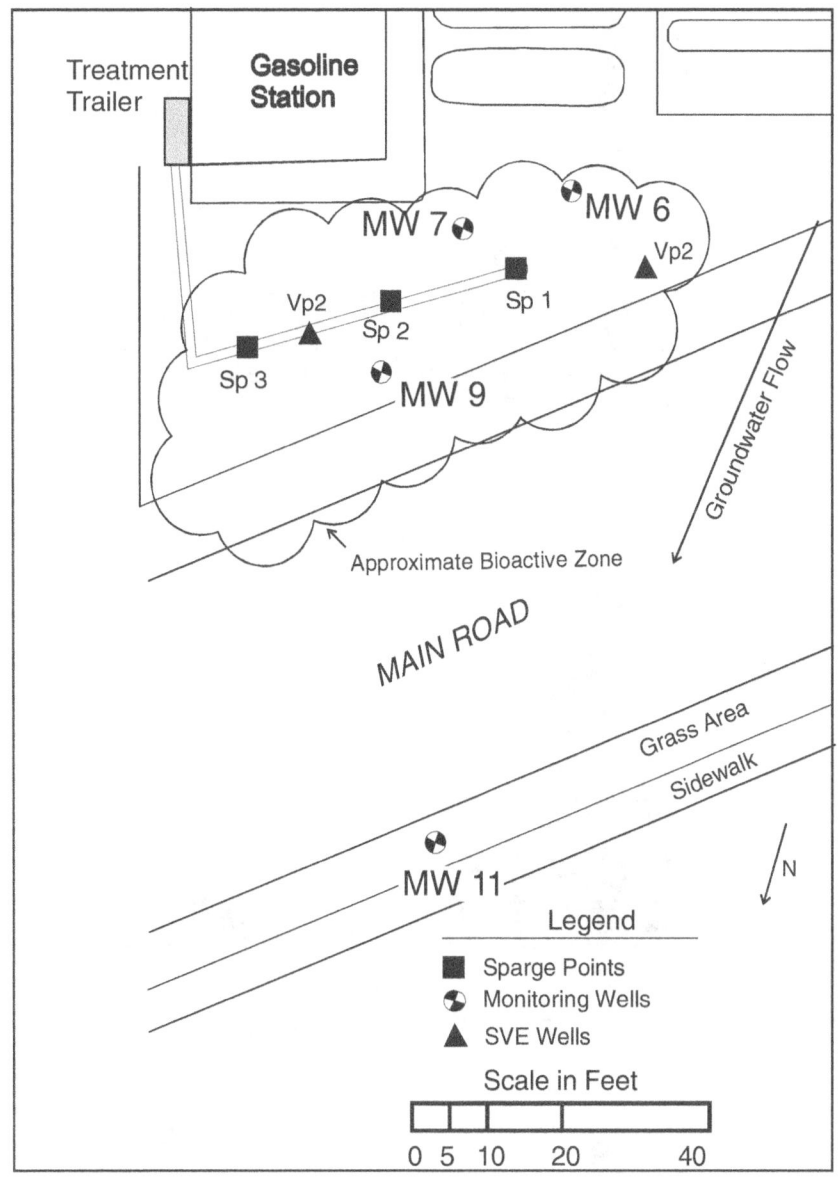

Figure A-1. Field site and system layout. Propane and air were injected into three existing air sparging points (Sp1, Sp2, and Sp3), and MTBE concentrations were measured in MW6, MW7, MW9, and MW11.

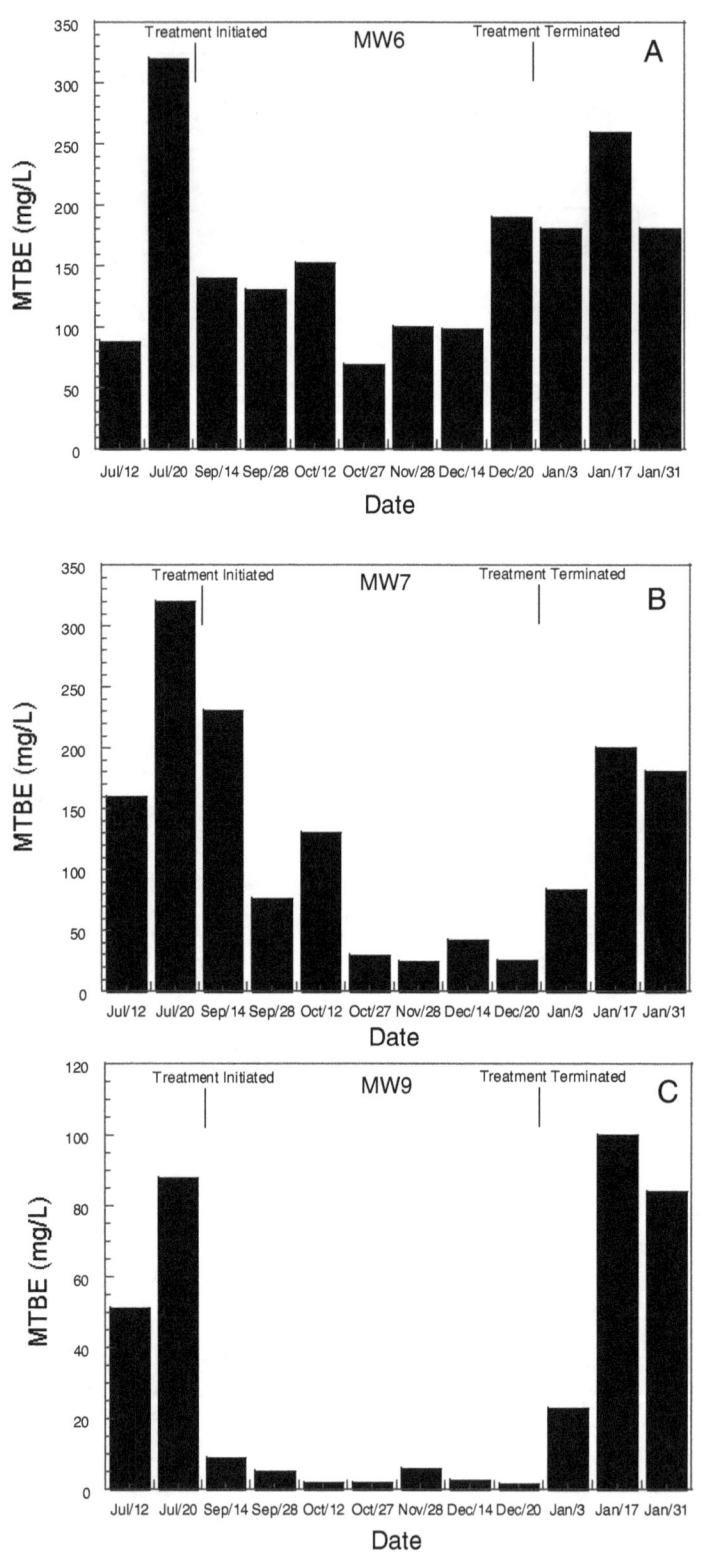

Figure A-2. MTBE concentrations in groundwater at on-site monitoring wells at a Camden County, New Jersey service station before, during, and after propane biostimulation treatment.

This corresponded to MTBE half-lives of 82, 24, and 30 days, respectively. After nearly 5 months of operation the treatment system was shut down. In each of the treatment zone monitoring wells the MTBE concentration rebounded to near pre-treatment levels (see Figure A-2A-C). The rebound effect was attributed to a continuing source of MTBE contamination at the site. Ongoing work at the site has led to a repair of the leakage source and implementation of an expanded treatment system for full-scale remediation of the site, including the source area.

TBA concentrations in the site groundwater increased during MTBE biodegradation, but they were typically several orders of magnitude lower than MTBE concentrations. During our initial work with propane oxidizing bacteria, pure cultures produced nearly stoichiometric concentrations of TBA from MTBE (Steffan et al., 1997). TBA concentrations in the cultures decreased only after MTBE was completely degraded. At this site, however, TBA was apparently degraded simultaneously with MTBE because it did not accumulate to levels near the initial MTBE concentration. Furthermore, TBA concentrations declined rapidly after propane injection was terminated and MTBE degradation ceased. The decline in TBA concentrations was accompanied by a decline in oxygen concentration. These data suggest that the propane oxidizers continued to degrade TBA after propane was no longer available to induce MTBE degradation, or that other TBA degraders were present in the system. During microcosm studies with ENV425 the organisms degraded TBA to <5 μg/L, indicating that similar levels will be achieved in the field provided the treatment period is sufficiently long.

The results of the case study showed that MTBE-contaminated groundwater can be biologically remediated using propane oxidizing bacteria and propane biosparging. This site presented a number of unique challenges to this technology, including low pH, high MTBE concentrations, and a continuing source of MTBE. Nonetheless, a significant mass of MTBE was removed during this demonstration, and MTBE reductions of greater than 90% were achieved in a relatively short time. The results also suggest that this treatment approach also supports the degradation of TBA.

Propane proved to be an excellent substrate for biostimulation applications; it is widely available, transportable even to remote sites, and relatively inexpensive. Application of propane injection in the field, however, may raise concerns about creating explosive mixtures of propane and air in situ. To address these concerns we injected propane in pulses and did not exceed 10% of the LEL of propane in the injection gas. We also used SVE to prevent in situ accumulation of propane. The results of our monitoring, however, suggest that propane stripping is minimal and SVE is likely unnecessary at most sites.

Technology Costs. Estimates of the cost of implementing the propane biostimulation system are similar to the costs of applying conventional air sparging/biosparging at a service station site. During the case study, propane costs were only $240 for the entire 6 months of operation. The primary equipment cost for the application is a biosparging system that safely blends low levels of propane with sparging air. A typical system, fully engineered, constructed and mounted in a trailer is expected to costs approximately $35,000, but the mobile system is suitable for repeated use at multiple sites, or it could be returned to a site to remediate future MTBE releases. Stationary systems can be installed at a lower cost. Based on the results of the project, future applications of the technology probably will not require the use of SVE during biosparging, saving both the equipment and discharge permit costs. It also is recommended that pre-design treatability studies be performed with site groundwater and soil. These tests are expected to cost ~ $4,000. Addition of seed cultures, when needed, is expected to cost ~$1000 to $2000 per application depending on the size of the site. The technology also can be applied in a number of alternative configurations — some employing existing systems — depending on site characteristics and treatment needs. Thus, the complexity of the site and the selection of an application design will ultimately determine the total cost of the system.

A.5   Summary

Propane biostimulation is a useful and economical in situ treatment alternative for remediating MTBE contaminated groundwaters. The technology is very flexible and can be combined with other traditional technologies like air sparging and soil vapor extraction to enhance the removal of MTBE from groundwater. Importantly, the technology also promotes the removal of TBA from groundwater. Because TBA is highly water soluble and not easily removed by air sparging, soil vapor extraction, or carbon adsorption, the ability to simultaneously remove MTBE and TBA in a single treatment process, and in situ, should present a considerable cost savings to users of the technology. Demonstrations performed to date show that the technology can be applied safely with little risk of fugitive propane emissions or accumulation in the subsurface.

Propane biostimulation should be considered as a remedial alternative for sites where air sparging or the addition of oxygen alone does not support MTBE remediation (see Case Study above). Likewise, it should be considered in regions of the country where TBA in groundwater also is tightly regulated. Furthermore, the potential application of propane biostimulation should be considered when installing an air sparging system at an MTBE contaminated site. By creating a flexible system that will allow the subsequent application of propane injection in the event that air sparging alone is not sufficient, considerable cost savings can potentially be realized in overall treatment costs. Similarly, the subsequent addition of propane for in situ biostimulation should be considered when planning the use of other technologies such as cut-off trenches and bioaugmentation with MTBE degrading microbes. In all cases, it is recommended that treatability studies be performed prior to designing and implementing propane biostimulation systems. Treatability studies can provide information about the availability of indigenous MTBE-degrading propane oxidizing microorganisms and provide insight regarding propane and oxygen loading requirements and the presence of geochemical conditions that could limit microbial activity (e.g., low pH).

A. 6   References Cited and Additional Suggested Information

Steffan, R. J., McClay, K., Vainberg, S., Condee, C. W., and Zhang, D. 1997. Biodegradation of the Gasoline Oxygenates Methyl *tert*-butyl ether, Ethyl *tert*-butyl ether, and *tert*-Amyl ether by Propane-Oxidizing Bacteria. Appl. Environ. Microbiol. 63: 4216-4222.

Steffan, R. J., Y. Farhan, C. W. Condee, and S. Drew. 2002. Bioremediation at a New Jersey Site Using Propane-Oxidizing Bacteria. In: Ellen Moyer and Paul Kostecki (Eds.), Chapter 27: MTBE Remediation Handbook, Amherst Scientific Publisher, Amherst, MA (in press).

Steffan, R. J., P.B. Hatzinger, Y. Farhan, and S. Drew. 2001. In Situ and Ex Situ Biodegradation of MTBE and TBA in Contaminated Groundwater. In Proceedings of the 2001 Petroleum Hydrocarbons and Organic Chemicals in Ground Water: Prevention, Detection, and Remediation Convention and Exposition, Nov. 14-16, Houston, TX, The National Groundwater Association, Westerville, OH, pp.252-264.

Steffan, R.J., P.B. Hatzinger, Y. Farhan, and S.R. Drew. 2001. In Situ and Ex Situ Biodegradation of MTBE and TBA in Contaminated Groundwater. In Proceedings of the National Groundwater Association Conference on Petroleum Hydrocarbons and Organic Chemicals in Ground Water: Prevention, Detection, and Remediation. Nov. 14-16, Houston, TX, pp.252-264.

Steffan, R. J., C. Condee, J. Quinnan, M. Walsh, S. H. Abrams, and J. Flanders. 2000. In Situ Application of Propane Sparging for MTBE Bioremediation. In G.B. Wickramanayake, A. R. Gavaskar, B.C. Alleman, and V. S. Magar (Eds.), Bioremediation and Phytoremediation of Chlorinated and Recalcitrant Compounds. Battelle Press, Columbus, OH, pp.157-165.

Steffan, R. J., S. Vainberg, C. W. Condee, K. McClay and P. Hatzinger. 2000. Biotreatment of MTBE with a New Bacterial Isolate. In G.B. Wickramanayake, A. R. Gavaskar, B.C. Alleman, and V. S. Magar, (Eds.), Bioremediation and Phytoremediation of Chlorinated and Recalcitrant Compounds. Battelle Press, Columbus, OH, pp.165-173.

Vainberg, S, A. P. Togna, P. M. Sutton, and R.J. Steffan. 2002. Treatment of MTBE-Contaminated Water in a Fluid Bed Bioreactor (FBR). J. Environ. Engineer 128:842-851.

www.ingramcontent.com/pod-product-compliance
Lightning Source LLC
Chambersburg PA
CBHW080640180526
45168CB00008B/3240